Toward a Directionalist
Theory of Space

Toposophia
Sustainability, Dwelling, Design

Toposophia is a book series dedicated to the interdisciplinary and transdisciplinary study of place. Authors in the series attempt to engage a geographical turn in their research, emphasizing the spatial component, as well as the philosophical turn, raising questions both reflectively and critically.

Series Editors:
Robert Mugerauer, University of Washington
Brian Treanor, Loyola Marymount University

Editorial Board:
Edmunds Bunkse, Kim Dovey, Nader El-Bizri, Matti Itkonen, Eduardo Mendieta, John Murungi, John Pickles, Ingrid Leman Stefanovic

Books in the Series:
Toward a Directionalist Theory of Space: On Going Nowhere, by H. Scott Hestevold
Urbanizing Carescapes of Hong Kong: Two Systems, One City, by Shu-Mei Huang
Mapping and Charting in Early Modern England and France: Power, Patronage, and Production, by Christine Petto
Remembering Places: A Phenomenological Study of the Relationship between Memory and Place, by Janet Donohoe
Spoil Island: Reading the Makeshift Archipelago, by Charlie Hailey
Reading the Islamic City: Discursive Practices and Legal Judgment, by Akel Isma'il Kahera
Metamorphoses of the Zoo: Animal Encounter after Noah, Edited by Ralph R. Acampora
The Timespace of Human Activity: On Performance, Society, and History as Indeterminate Teleological Events, by Theodore R. Schatzki
Environmental Dilemmas: Ethical Decision Making, by Robert Mugerauer and Lynne Manzo
When France Was King of Cartography: The Patronage and Production of Maps in Early Modern France, by Christine Petto
Mysticism and Architecture: Wittgenstein and the Meanings of the Palais Stonborough, by Roger Paden

Toward a Directionalist Theory of Space

On Going Nowhere

H. Scott Hestevold

LEXINGTON BOOKS
Lanham • Boulder • New York • London

Published by Lexington Books
An imprint of The Rowman & Littlefield Publishing Group, Inc.
4501 Forbes Boulevard, Suite 200, Lanham, Maryland 20706
www.rowman.com

6 Tinworth Street, London SE11 5AL, United Kingdom

Copyright © 2020 The Rowman & Littlefield Publishing Group, Inc.

All rights reserved. No part of this book may be reproduced in any form or by any electronic or mechanical means, including information storage and retrieval systems, without written permission from the publisher, except by a reviewer who may quote passages in a review.

British Library Cataloguing in Publication Information Available

Library of Congress Cataloging-in-Publication Data Is Available

ISBN 978-1-4985-7996-4 (cloth)
ISBN 978-1-4985-7998-8 (pbk)
ISBN 978-1-4985-7997-1 (electronic)

To Nita, Erik, and Kara—members of a family that confirms that at least one whole is even greater than the sum of its parts.

Contents

Preface	ix
1 The Map to Nowhere and Beyond	1
2 Spatial Directionalism	17
3 A Directionalist Theory of Space	33
4 Defending Spacelessness	49
5 The Special Composition Question Revisited	83
6 Does the Road to Nowhere Include Boundaries and Holes?	113
7 Is Modern Physics a Roadblock to Going Nowhere?	137
Bibliography	191
Index	199
About the Author	205

Preface

Most of us believe that physical objects exist, but do there also exist the places *where* those objects exist or *could* exist? Does there exist something that *is* the place where your left hand is located, and do there exist other things that *are* the places where your left hand *could* be located should you extend your arm to the left or to the right to gesture toward, respectively, Mount Rainier or Mount Hood? And, if you do gesture toward a mountain, does the extending of your arm involve your hand's occupying successively different *places* at different times? Given Earth's rotation, does Mount Rainier occupy now the same place that it occupied an hour ago . . . or a second ago? Is the place occupied by your left hand smaller than the place occupied by Mount Rainier but larger than the place occupied by one of your left hand's protons?

Isaac Newton argued that *places* exist and are regions of (absolute/substantivalist) space that may or may not be occupied. Newton's contemporary, Gottfried Wilhelm Leibniz, argued instead that places do *not* exist and that talk about occupied and unoccupied regions of space should be construed as talk about the spatial *relations* that protons, hands, mountains, planets, and stars bear or *could* bear to one another.

After a brief review of arguments for and against Newton's substantivalism and Leibniz's relationalism (chapter 1), I entertain the view that there is a class of spatial relations that Leibniz overlooked—spatial *directional* relations (chapter 2)—and I then formulate a new relationalist theory of space in terms of these relations (chapter 3). After explaining how the new relationalist theory can itself preserve the plausible Newtonian intuitions that motivate standard arguments on behalf of absolute/substantivalist *space* (chapter 4), I invoke spatial directional relations to reformulate my original theory of composition—my original theory of when it is that nonoverlapping objects compose, and are composed by, other objects (chapter 5). I also invoke spatial directional relations to sketch reductivist theories of boundaries and holes (chapter 6). Finally, I bring this study to a close by extending the new relationalist theory of location to the nature of *spacetime* (chapter 7). Consider first, however, a bit of background regarding my interest in objects, composition, time, and space.

I began thinking about philosophy of objects as a first-year graduate student enrolled in Roderick Chisholm's Metaphysics. After he presented an amusing formulation of the problem of the Ship of Theseus, Chisholm entertained various solutions before he introduced students to mereology and then developed his own solution. Two years later, when I audited Metaphysics as a third-year student, Chisholm addressed again the Ship of Theseus; but, this time, to close the last of the classes devoted to objects and diachronic identity, he off-handedly remarked about "philosophy of contact" and offered a seemingly unplanned sketch of a problem involving touching. Noting that the concept of contact (i.e., touching) is problematic, Chisholm cleared his throat and quipped dryly, "This is the closest to philosophy of sex we'll get, and the result may be disturbing."[1] First, Chisholm observed that two discrete (i.e., nonoverlapping) things cannot be in exactly the same place at the same time. So, if a cube's left half is in contact with its right half, their touching could *not* involve the left half-cube's having a tiny part (e.g., a surface or "outer boundary") that occupies the same place as a tiny part of the right half-cube. But, if no part of the left half-cube occupies the same place as a part of the right half-cube, then the half-cubes could not be in contact—they would be separated and thereby could not be touching each other. The class ended with the formulation of this puzzle and the "disturbing" suggestion that touching may be impossible.[2] During the next class meeting, Chisholm began addressing problems involving personal identity and said nothing more about contact; but I continued to contemplate contact and then composition: If touching *is* impossible, then *is* it possible that there exists a cube composed of two half-cubes if the half-cubes' surfaces ("outer boundaries") cannot be in contact? If two half-cubes *do* compose a cube, would the cube cease to exist if the half-cubes' surfaces cease to touch? *Does* composition involve one object's surface being *in contact* with another object's surface? But what *is* a surface, and in what sense can two discrete surfaces be *in contact*? I soon decided that this nest of questions would constitute the topic of my dissertation.

Regarding the problem of contact, Chisholm referred me to Brentano's work on spatial and temporal boundaries. Initially, I was skeptical of Brentano's view that a "materially solid" object would be a "continuum" composed of parts that include non-three-dimensional "inner" and "outer" boundaries. Eventually, I concluded that Brentano was right that a "solid" cube *would* be a "material continuum" that has a two-dimensional square part that would be the "inner" boundary that, in some sense, lies *between* the cube's composing half-cubes. But I also became convinced that composition could not be explained solely in terms of the "touching" or "overlapping" of the "inner" boundaries of an object's composing half-cubes—that an adequate theory of composition should allow for both "solid" objects *without* "gaps" among composing parts *and* for "scattered" objects *with* "gaps" among at least some of their composing parts. I sought a theory of composition that would allow for the possibility that a hydrogen atom is an object composed of a one-proton, one-neutron nucleus *and* an electron even though the nucleus and electron are separated by a "gap"—even though the nucleus and electron do not *touch*.

Ultimately, in my 1978 dissertation and subsequent articles, I defended both Conjoining—the view that two objects compose a whole when no third object lies between them—and Brentano's view that "materially solid" objects are "continuous wholes" that have "inner" and "outer" boundaries that are non-three-dimensional parts.[3] I did, however, resist Brentano's view that discrete boundaries can "coincide" ("occupy the same place").

My study of diachronic identity and objects led, understandably, to my study of time. In an extended article sympathetic with Presentism ("If x exists, then x presently exists"), I argued that a Presentist should side with Leibniz and resist commitment to substantivalist *times*, and I then offered a formulation of Presentism cast in terms of temporal relations without reference to temporal locations.[4] This work on temporal relations and location led naturally to my study of the nature of spatial location and to my again taking seriously Leibniz's relationalism. The result is this formulation of a new relationalist theory of space and a study of its implications for the Special Composition Question, the nature of boundaries and holes, and the nature of spacetime.

I remain grateful to many who have helped me with this project over the past several years. First, I thank physicist Patrick LeClair. When I began searching for a better understanding of how to connect philosophical work on time and space with the physics of time, space, and spacetime, I asked the chair of UA's Department of Physics and Astronomy for advice, and he referred me to LeClair's class, Modern Physics. I audited Modern Physics in 2010; and, during one of our after-class conversations, LeClair suggested that we co-teach a course on topics at the intersection of physics and metaphysics. In 2012 and 2014, we co-taught Physics, Metaphysics, and Other Nonsense, and, each time, this course attracted a couple dozen wonderful students—half physics students, half philosophy students. To complement LeClair's remarkable non-mathematical explication of General Relativity, I addressed the Newton/Leibniz dispute and then produced a handout that allowed me to present a preliminary sketch of a new directionalist/relationalist theory of space that I had been contemplating for at least a year. This handout eventually grew into an early draft of chapters 2 and 3. For the many questions, comments, and objections that helped shape this work, I thank LeClair and the students in our two "Nonsense" classes. I am also grateful to students in my 2018 seminar Space, Time, and Objects—students with whom I have more recently discussed the metaphysics of temporal and spatial location; my exchanges with Julie Jackson were particularly helpful.

I am especially grateful to physicist LeClair and his colleague, astronomer Raymond E. White, for innumerable conversations over the past five years. I thank them for addressing patiently a metaphysician's questions and conjectures; and I continue to admire their scientific expertise *and* philosophical sensitivity. If this book contains errors regarding Relativity and spacetime, LeClair and White are most certainly *not* to blame.

I thank The University of Alabama and its College of Arts and Sciences for the sabbatical leave that allowed me to finish a preliminary draft of chapters 1–4 and working outlines for chapters 5–6. And, I thank the University, the College, the Department of Physics and Astronomy, and the Department of Philosophy for providing me with a secluded office and research resources during the year that I completed the manuscript. I also thank those alumni and friends of the Department of Philosophy who contributed to the departmental gift fund that can be used to support faculty research, and I thank my chair, Richard Richards, for making such support available. The generosity of the Department's alumni and friends has supported both travel to conferences *and* the production of diagrams for this book. For the diagrams themselves, I thank graphics designers Christina Frantom and Jeremy Rich of White Roche: they produced clear interpretations of my embarrassingly dreadful sketches.

My work on location has benefitted from presentations at the 2016 Mid-South Philosophy Conference, at a conference on "The Continuing Relevance of Leibniz" at Franciscan University, and at a departmental faculty colloquium. I thank colleagues Torin Alter, Holly Kantin, Norvin Richards, and Richard Richards for their helpful comments; and I am especially grateful to Tim Butzer and Chase Wrenn for their particularly penetrating objections. For useful suggestions and references, I thank a reviewer who, to me, remains anonymous. I thank my son, Erik Hestevold, for enjoyable conversations about space and spacetime and for producing the painting ("Multiple Directional Infinities") that appears on this book's cover. My son's philosophical mind and artist's eye have allowed him to both understand *and* represent a directionalist theory of location. I am also grateful to my wife, Nita DeBoer Hestevold, for her encouragement, for her philosophical observations, and for her patience during our proofreading and indexing of the proofs.

Finally, I thank Emory Kimbrough whom I met several decades ago through a common interest in conjuring. With undergraduate work in mathematics, physics, and philosophy and with graduate work in high-energy physics, Kimbrough is a professional magician and juggler who is also an invaluable interlocutor regarding the interface between physics and metaphysics. I thank Kimbrough for thirty years of conversations about physics and philosophical problems involving time and space, and I thank him for reading every word of my penultimate draft. I remain grateful for his innumerable comments, suggestions, corrections, and objections.

NOTES

1. My class notes indicate that, on November 8, 1976, Chisholm devoted most of his Metaphysics class to formulating a final objection to "temporal parts" theories of diachronic identity. Not until the last few minutes of class did he mention casually the problem of contact. This was the first time that I had heard Chisholm mention this problem; and I later concluded that his interest had been motivated by his study of Franz Brentano's work on spatial and temporal boundaries.

2. In his review of what he claims are inconclusive arguments against points of space, Frank Arntzenius neatly formulates (but does not defend) a version of this argument for "[t]he impossibility of genuine contact"; see *Space, Time, and Stuff* (Oxford and New York: Oxford University Press, 2012), p. 133.

3. See, respectively, H. Scott Hestevold, "A Metaphysical Study of Aggregates and Continuous Wholes," PhD diss. (Brown University, June 1978); "Conjoining," *Philosophy and Phenomenological Research* 41 (1981), 371–85; and "Boundaries, Surfaces, and Continuous Wholes," *The Southern Journal of Philosophy* XXIV (1986), 235–45.

4. H. Scott Hestevold, "Presentism: Through Thick and Thin," *Pacific Philosophical Quarterly* 89 (2008), 330–31.

Chapter 1

The Map to Nowhere and Beyond

SPATIAL SUBSTANTIVALISM AND SPATIAL RELATIONALISM

When you hold a pen, how many entities are there between your fingers? Just the pen? Or does there exist both the pen *and* the *place* occupied by the pen? Is there a place occupied by the pen that is larger than the place occupied by the pen's cap? As you write a note with the pen, does the pen's movement involve the pen's occupying a different *place* at each successive moment of movement? When you place the pen on your desk after finishing the note, is there a place that it then occupies that is other than the place that it occupied when you held it between your fingers? When you leave the pen on your desk overnight, is the place that it occupies the following morning the same place that it occupied the night before?

The philosophical question at stake is this: In addition to pens and protons, dogs and daffodils, stars and stones, and whatever other objects there may be, do there also exist the *places*—the *locations*—that those objects occupy or *could* occupy?

Spatial Substantivalism

Isaac Newton argued that such places exist and that they are regions of *absolute space*—regions of a nonphysical "immovable" thing *in* which pens and other physical entities are located:

> Absolute space, in its own nature, without relation to anything external, remains always similar and immovable. . . . For if the earth . . . moves, a space of our air, which relatively and in respect of the earth, remains always the same, will at

one time be one part of the absolute space into which the air passes; at another time it will be another part of the same, and so, absolutely understood, it will be continually changed.[1]

Eighteenth-century Scottish philosopher Thomas Reid also endorsed the existence of *space*:

> [A] body could not exist if there was no space to contain it. It could not move if there was no space.
>
> We see no absurdity in supposing a body to be annihilated; but the space that contained it remains; and to suppose that annihilated seems absurd.
> We can set no limits to [space], either of extent or of duration. Hence we call it immense, eternal, immoveable, and indestructible.[2]

Newton and Reid would agree that objects occupy places in the sense that they are located in absolute space—that each object occupies a particular region of absolute space.[3] If you submerge your pen in a tall glass of water, your pen would not occupy *water*; rather, Newton would claim, the pen would occupy a region of absolute space that is the same region of space formerly occupied by the mass of water that the pen displaced.

Using current terminology, Newton, Reid, and others who believe that absolute space or regions of absolute space exist are *Spatial Substantivalists*:

> SS There exist regions of *space* [i.e., places, spatial locations].[4]

If regions of substantivalist space exist, they would be unperceivable, non-substantial particulars that are themselves without location. What would motivate one to believe that such bizarre particulars exist? The arguments for Spatial Substantivalism include the following:

- "Kidneys, dogs, rocks, and pens *do* have *locations*, and postulating the existence of regions of space is the only plausible way to make sense of facts involving locations."
- "Things *move*, and postulating the existence of regions of space is the only plausible way to make sense of facts involving movement: To move *is* to occupy different regions of space at different times."
- "'Here' and 'there' have referents, and particular regions of space are the only entities that could be the referents for these spatial indexicals."
- "Holes and crevices exist, but holes and crevices are nothing more than particular regions of space surrounded by matter—regions of space that are entirely empty or occupied by 'filler' (e.g. air, putty, dental amalgam)."

- "Postulating the existence of regions of space is the only plausible way to explain what metersticks do: They measure an object's extension *in space*."
- "Certain geometric facts imply that regions of space exist. The fact that there is a spot of dried coffee that covers the midpoint of a newspaper's front page implies that there exists a point-sized region of space that is the midpoint of the one-dimensional region of space that extends between *the* points occupied by, respectively, the front page's upper-right and lower-left corners."

Consider four more arguments for substantivalist space—arguments that will be addressed later in detail:

- "It is possible that *all* physical objects move uniformly—that they all move in the same direction at the same velocity without any alteration of the spatial relations that obtain among these objects. But what would such uniform movement involve? Given that there would be no perceivable change of spatial relations among these objects, their uniform movement must involve their successively occupying different regions of substantivalist space at each successive moment of movement."
- "There is a possible world that is the inverted version of the actual world: That is, the inverted world is exactly like the actual world except that the array of physical objects in the inverted world is the exact mirror-image of the array of physical objects in the actual world. But what, *exactly*, would the difference between the inverted world and the actual world involve? The difference could *not* involve a difference in spatial relations: If your pen and left hand are both to the left of your heart when you finish writing a note in the actual world, then they would also be to the left of your heart when you finish writing the note in the inverted world. Thus, the inverted spatial orientation of objects would involve those objects having an inverted orientation *in space*: The objects 'on one side of space' in the actual world would occupy 'the other side of space' in the inverted world."
- "It is possible that, at some given time, every spatially extended physical object doubles in size such that all such objects continue to bear the same spatial relations to other entities after the doubling. If your pen is one meterstick's length to the left of your desk lamp, it will remain a meterstick's length away if the pen, your lamp, the meterstick, and all other physical objects double in size. Such doubling, then, would *not* involve a change in spatial relations that obtain among the objects that double. Instead, such doubling would involve substantivalist space: The region of space that a meterstick would occupy after the doubling would be twice the size of the region that it occupies before the doubling."

- "There is a possible world W that includes no other physical object but spinning cannonball B; and there is another possible world W* that includes no other physical object but cannonball B at rest. The only difference between W and W* is that B *spins* in W but not in W*. But what does B's spinning involve? B's spinning cannot involve B's parts changing the spatial relations that they bear to other objects: In W, there exist *no* other objects relative to which B's parts could be changing their spatial orientation as B spins. Thus, B's spinning involves B's parts successively occupying different regions of substantivalist space as B spins."

In recent years, some philosophers have carefully left open the question of whether substantivalist *space* exists[5] while others have presupposed its existence. Friends of substantivalist space include Richard Cartwright and Ned Markosian:

> By a *region of space*, ... let us agree to understand any set of points of space. And by a *receptacle* let us understand a region of space with which it is possible some material object should be, in Hobbes' phrase, coincident or coextended.[6]
>
> I will assume without argument ... that physical objects are objects with *spatial locations*.
>
> ...
>
> The mereological properties and relations of physical objects are determined by the mereological properties and relations of the spatial regions those objects occupy.[7]

Consider an alternative to positing the existence of ethereal "absolute" *space*.

Spatial Relationalism

Leibniz adamantly rejected substantivalist space, claiming that an object's location is not a region of *space*, but is instead nothing more than the sum of those spatial *relations* that the object bears to other objects:

> I hold space to be something merely relative as time is; that I hold it to be an order of coexistence, as time is an order of successions. For space denotes, in terms of possibility, an order of things which exist at the same time.
>
> ...
>
> I will here show, how men come to form to themselves the notion of space. They consider that many things exist at once and they observe in them a certain order of co-existence, according to which the relation of one thing to another is more or less simple. This order, is their *situation* or distance. When it happens that one of those co-existent things changes its relations to a multitude of others, which

do not change their relation among themselves; and that another thing, newly come, acquires the same relation to the others, as the former had; we then say, it is some into the place of the former; and this change we call motion.[8]

The passages above suggest that Leibniz would endorse *Spatial Relationalism*:

> SR It is false that *space* or regions of *space* exist; reasonable claims that imply that *space* or regions of *space* exist can be reformulated as claims involving spatial entities or spatial relations that clearly do not imply that *space* or regions of *space* exist.[9]

Leibnizian Spatial Relationalists would insist that the present location of one's pen is not a particular region of *space*, but is instead those spatial relations that the pen presently bears to other such spatial entities: The pen's being a certain number of centimeters at a certain angle below one's nose, *and* the desk lamp's being a certain number of centimeters at a certain angle above one's pen, and so on. To move the pen from one's hand to one's desk is *not* to bring it about that the pen occupies successively different regions of space; rather to move the pen is to bring it about that the spatial relations that the pen in the hand bears to one's nose and the lamp is other than the relations that the pen on the desk bears to one's nose and the lamp.[10]

Leibniz himself makes clear that Spatial Relationalism is a reductivist theory of space: He compares (a) the reduction of *place*-talk to spatial-relation-talk with (b) the reduction of talk of *place*-in-a-*family*-talk to *family*-relation talk.[11] Strictly, there exists nothing that is someone's *family* such that the individual family member occupies a particular *place* within that family. Rather, to say that Elsie occupies the places of both daughter and mother in the Bryant family is to say that Elsie bears different familial relations to other people: There exist at least two people such that Elsie is the daughter of one and the mother of the other. Similarly, then, to talk of the *place* (or *region of space*) that a pen occupies is to talk of the various spatial relations that the pen bears to other bearers of spatial relations.

The obvious defense of Spatial Relationalism is its ontological simplicity: Whereas Spatial Substantivalism implies that there exist both regions of space *and* spatial relations; SR implies that there exist spatial relations but *no* regions of space. Some will insist that there is good reason to endorse the larger ontology—that only by postulating the existence of "unperceivable," "eternal, immoveable, and indestructible" space can one preserve the distinction between absolute and relative motion. And, they will continue, only by positing absolute (substantivalist) space can one preserve the possibilities that all objects move uniformly, that the inverted version of the actual world could have existed instead, that all objects double in size, and that a cannonball can spin in the stark world in which it and its parts are the only objects.

In chapters that follow, I take seriously Leibniz's effort to lead us down the road to no*where*—down the road to a theory of *space*lessness. By invoking certain *direction*-involving spatial relations that Leibniz overlooked, I will formulate a new *directionalist* version of Spatial Relationalism that enjoys the ontological economy of Leibniz's view while preserving the absolute motion central to Newton's view.

According to Leibniz's version of Spatial Relationalism, it is impossible that all objects move uniformly because there exists no substantivalist space relative to which objects move *and* there can be *no* relational difference between the world in which all objects are at rest and the world in which all objects move uniformly. According to the new *directionalist* version of Spatial Relationalism, although there exists no substantivalist space relative to which objects move, moving objects *do* undergo a change of *directional* relations that they exhibit. Similarly, without substantivalist space, the *Leibnizian* Spatial Relationalist should agree that there is no possible world that is the inverted version of the actual world, that it is impossible that all objects double in size, and that it is impossible that a cannonball spins in a world in which it and its parts are the only objects. The *directionalist* Spatial Relationalist should claim instead that, although there exists no substantivalist space relative to which objects could be inverted, double in size, or spin, the objects in the inverted world, objects that have doubled in size, and the solitary spinning cannonball would all exhibit different *directional* relations than, respectively, objects in the actual (noninverted) world, objects before doubling in size, and a cannonball at rest.

THE DIRECTIONALIST PROJECT

In chapter 2, after identifying types of spatial (i.e., spatially locatable) entities that *may* exist, I explicate *Spatial Directionalism*—the view that the universe is (absolutely) *directioned* such that, in addition to the spatial relations that Leibniz countenanced, there also exist spatial *directional* relations that obtain among objects and any other spatial entities that exist. And, the concept of a *spatial entity*—"something that has a spatial location"—is explicated in terms of spatial directional relations.

In chapter 3, I appeal to spatial directional relations to analyze the concept of a spatial entity's *possible location*—a (spatial) location that is (or could be) the location of that spatial entity. I then use the concept of a *possible location* to analyze directionally/relationally the concept of a spatial entity's *actual location*, and I use the directionalist/relationalist concept of an actual location to analyze the mereological concepts of a proper part, of nonoverlapping spatial entities, and of composition. The directionalist/relationalist account of an *actual location* allows a precise formulation of the *Directionalist Theory of Space*—a version of Spatial Relationalism that, unlike Leibniz' version, *can* preserve absolutist/substantivalist intuitions about uniform movement, the inverted world, and absolute/relative motion. Finally, after noting that the Directionalist Theory of Space is consistent with the existence of more than three *spatial* dimensions, I offer a directionalist/relationalist account of spatial dimensionality.

In chapter 4, I address the Spatial Directionalist's responses to five defenses of Spatial Substantivalism: four classic arguments *for* substantivalist *space* plus an objection to spatial directional relations themselves. The four classic arguments turn on four plausible presuppositions: That it is possible that all spatial entities move uniformly (while no entity changes the spatial relations that it bears to other entities), that the inverted, mirror-image version of the actual world could have existed instead (such that the spatial relations that obtain between any two entities in the actual world would be the same that obtain between those two entities in the inverted world), that it is possible that all objects uniformly double or triple in size (while no object changes the spatial relations that it bears to other objects), and that, in a world otherwise devoid of spatial entities, a bucket of water or bola could exhibit motion-caused forces even though the object could not move relative to others (given that there would exist no other spatial entities relative to which the bucket or bola could change spatial relations). The fifth defense of Spatial Substantivalism involves this objection to spatial directional relations themselves: "Positing spatial directional relations is dubious because there is *no* criterion of identity for such relations!"

With respect to the four classic arguments for SS, the *Leibnizian* Spatial Relationalist would flatly reject the plausible intuitions on which these arguments rest, treating them as casualties of embracing the ontologically simpler relationalist theory of space. And, the Leibnizian would altogether ignore the fifth defense of SS: The fifth defense is an objection to spatial directional relations themselves, and the Leibnizian version of Spatial Relationalism does not imply the existence of such relations.

With respect to the four classic arguments for substantivalist *space*, I argue that, by countenancing spatial directional relations, the Spatial Directionalist can object to the arguments *without* rejecting the arguments' plausible premises involving the possibilities of uniform movement, spatial orientation, uniform expansion, and motion-caused forces. Whether the Spatial Directionalist's objections are successful will be a function, of course, of whether the directionalist can satisfy the critic who demands a criterion of identity for spatial directional relations. After entertaining several directionalist responses to the critic (including the formulation of a criterion of identity for spatial directional relations), I end the chapter with a modest defense of the Directionalist Theory of Space.

In chapter 5, I revisit my original answer to a question that I raised in a 1981 article—the question now known as "the Special Composition Question" [SCQ]: If there exist two discrete objects that do not compose a whole, what must happen to bring it about that they *do* compose a whole? Or, put another way, if there does exist a whole composed of two discrete objects, what must happen to bring it about that those two objects exist but cease to compose a whole? My original answer is that two objects compose a whole when *conjoined*—when no other nonoverlapping object lies between them.[12] This "conjoining" answer allows both that there *can* exist two nonoverlapping objects that do *not* compose a whole and that there *can* exist wholes that are bacteria, brains, humans, trees, pens, boulders, and stars. The answer also allows for the possibility that at least some wholes have parts that are *mereological simples*—zero-dimensional, *part*-less "atoms."

One objection to my original answer to SCQ (and to others' answers that followed) is that it suffers circularity—that it involves presupposing necessary and sufficient conditions for composition in order to specify the necessary and sufficient conditions for composition. This is among the objections that have motivated many to embrace extreme answers to SCQ: Some have argued that *any* two objects (e.g., your liver and the topmost stone of the Great Pyramid of Khufu) compose a whole; others have denied that the existence of objects are atomic nuclei, vital organs, stones, lamps, and stars, endorsing the nihilistic view that *no* two objects compose a whole. In chapter 5, to sidestep the charge of circularity, I appeal to directionalist analyses of fundamental mereological concepts to reformulate my original "conjoining" answer to SCQ. The chapter ends with my replies to several additional objections to "conjoining" and with my objections to two alternative answers to SCQ.

In chapter 6, I address briefly the metaphysical problems involving boundaries and holes: Is there a part of an object that is the object's "outer surface?" If so, how thick is that part? If there exists a materially solid sphere, is there a circular "inner" boundary in the middle? If so, is it part of the left hemisphere? The right hemisphere? Both hemispheres? Neither? Is it possible to strip away from the sphere nothing more than the sphere's surface? Is there a part of a doughnut or tire that is the *hole* in that object? If so, of what is the hole composed? If one fills a hole in a wall with putty, would the wall still have a hole in it?

Some have argued that boundaries are essentially (ontologically) dependent particulars—that they cannot possibly exist without the existence of a "host" object. For example, Franz Brentano claimed that "inner" and "outer" boundaries are ontologically dependent in the sense that they are essentially *non*-three-dimensional constituents of *some* three-dimensional materially solid whole or the other.[13] Brentano would claim that the materially solid sphere has a two-dimensional circular constituent that is the middle "inner" boundary that "separates" the sphere's hemispheres; and, Brentano would also claim that a materially solid cube would have a two-dimensional square top surface as well as a dozen one-dimensional "outer" boundaries that are the cube's edges. With respect to boundaries, then, Brentano would count as a nonreductivist: He would agree that reasonable claims that imply that non-three-dimensional boundaries exist *cannot* be reformulated as claims that clearly do not imply the existence of such ontologically dependent constituents.

Regarding holes, some may defend a reductivist theory, identifying an object's hole with a particular region of empty substantivalist *space*. Other reductivists may identify holes with certain physical entities—*hole-linings*—that are proper parts of holey physical objects.[14] There are those, however, who are *non*reductivists, claiming that holes, dents, and tunnels are metaphysically strange entities that are "parasitic" on the extended objects of which they are ontologically dependent constituents.[15]

I formulate in chapter 6 directionalist accounts of boundaries and holes cast in terms of the directionalist/relationalist concept of a location. If these accounts are successful, then endorsing Spatial Directionalism may afford one not only the ontological economy of a reductivist theory of space but the ontological economy of reductivist theories of boundaries and holes.

In chapter 7, I address the interface between the Directionalist Theory of *Space* and findings of modern physics. Do findings of modern physics provide adequate evidence that substantivalist *spacetime* exists? If so, does this count against relationalist theories of location? Or, by endorsing *spatiotemporal* directional relations, could one develop a plausible relationalist theory of spacetime, reducing *spacetime*-talk to talk about spatiotemporal entities and the spatiotemporal directional relations that obtain among them?

After formulating Spatiotemporal Substantivalism and also "the favored" interpretation of Special and General Relativity vis-à-vis spacetime, I appeal to *spatiotemporal* directional relations to formulate a directionalist/relationalist theory of *spacetime*. I then explicate five reasons why "the favored" interpretation invokes Spatiotemporal Substantivalism: It appears to allow for plausible accounts of absolute motion, fields (e.g., the electromagnetic field), gravity, the expansion of the universe, and gravitational waves. After formulating reasons as to why a metaphysician may be skeptical about the existence of substantivalist spacetime, I ultimately leave open the question of whether substantivalist spacetime exists and explain why the Directionalist Theory of *Space* may nonetheless be philosophically significant even if physicists and philosophers of physics offer convincing evidence that regions of substantivalist spacetime *do* exist.

WHAT SPATIAL ENTITIES MAY THERE BE?

The directionalist/relationalist theory of space to be developed in chapter 3 commits one to the existence of nothing other than spatial relations (including spatial *directional* relations), sets (or properties), states of affairs, and *spatial entities*—those entities among which spatial relations could obtain. But what spatial entities might there be? Exactly, what sorts of entities *could* count as spatial entities? Though the Directionalist Theory of Space does not itself imply *which* spatial entities exist, consider the candidates—a matter that I will address more formally in chapter 5.

Materially solid entities. If they do exist, three-dimensional, *materially solid* physical objects would count as spatial entities. Though I assume that every three-dimensional materially solid entity would have proper parts,[16] no materially solid entity could possibly be composed of just two "scattered" proper parts that are "entirely separated" by a "gap" between them. Rather, for any two proper parts of a materially solid entity, either those two parts are "directly contiguous" ("in contact") such that no "gap" and no other proper part "lies between" them, or those two parts are each "directly contiguous" with a third proper part that *does* "lie between" them. One should allow, however, that there exist some materially solid objects that have holes or gaps. For example, one should allow for the possibility that there exist *materially solid* doughnut-shaped or V-shaped objects. The C-shaped left half of a materially solid doughnut-shaped object would have proper parts that are "directly contiguous with" (i.e., *not* separated by "gaps" or by any other proper part from) certain proper parts of the doughnut-shaped object's C-shaped right half. (The left half would also have a proper part that is "directly across the hole" from a proper part of the right half; although these proper parts *would* be "separated by the hole," they would each be "directly contiguous" with a third materially solid proper part that "lies between" them.) Though I do presuppose that all three-dimensional materially solid spatial entities would have three-dimensional materially solid proper parts, I take no stand on whether it is necessarily true that every proper part of a three-dimensional materially solid entity does itself have proper parts. That is, I leave open the question of whether the proper parts of a three-dimensional materially solid entity would include zero-dimensional *mereological simples*.[17]

Mereological simples. I assume that, if *mereological simples* ("mereological atoms," "simplons") exist, they would be *zero-dimensional*, point-sized, *part*-less entities that would count as spatial entities. Though at least one spatial relation would obtain "externally" between any two mereological simples, no mereological simple could exhibit a spatial relation *internally*: Given that mereological simples would have no parts at all, no spatial relation could possibly obtain *between* two proper parts of a mereological simple—simples would be essentially *proper-part*-less. (If, for example, electrons are zero-dimensional particles, then electrons would count as mereological simples.) In subsequent chapters, I leave open the possibility that mereological simples exist and are parts of three-dimensional materially solid spatial entities, and I also leave open the possibility that mereological simples exist and are parts of *scattered objects*.

Scattered objects. Scattered objects would be objects that have at least two proper parts that are *not* "directly contiguous" and are "entirely separated" by "gaps."[18] *If* a hydrogen atom *is* an object composed of a nucleus around which a single electron orbits, then the hydrogen atom would not be a materially solid whole composed of "directly contiguous" proper parts; rather, the hydrogen atom would be a *scattered* object given that its nucleus and electron would be "*non*-directly-contiguous" proper parts "entirely separated" by a "gap." The directionalist/relationalist theory of space (developed in chapter 3) and my revised answer to the Special Composition Question (developed in chapter 5) both leave open the questions of whether scattered objects exist and whether scattered objects can have parts that are mereological simples.

Boundaries. I also leave open the possibility that "inner" and "outer" *boundaries* are among the spatial entities that exist—that there exist non-three-dimensional constituents of materially solid objects that can be bearers of spatial relations. For example, if a materially solid cube's constituents include six square two-dimensional "outer" boundaries (i.e., the cube's faces), then I leave open the possibility that spatial relations could obtain between any two of these "outer" boundaries and also between any one of these "outer" boundaries and any other nonoverlapping spatial entity. The directionalist/relationalist theory of space to be developed leaves open the question of whether boundaries exist, but in chapter 6, I suggest skepticism with respect to such *non*reductivist boundaries and defend instead a directionalist/reductivist theory of boundaries.

Also left open is the possibility that a zero-dimensional "outer" boundary exists and is one of a materially solid cube's corners while another zero-dimensional "inner" boundary exists and is "*the* center" of the cube. Presumably, there would be a significant metaphysical difference between a zero-dimensional "outer" or "inner" boundary and a mereological simple. Brentano and his defenders would claim that boundaries are dependent particulars—spatial entities that are essentially parts of some other spatial entity. Zero-dimensional mereological simples, however, would *not* be dependent particulars; any given "free-standing" simple could exist without being a part of anything else.

Monads. There would also be a significant metaphysical difference between a *part*-less entity that is a mereological simple or zero-dimensional boundary and a *part*-less entity that is a Leibnizian monad: Unlike mere mereological simples and zero-dimensional boundaries, monads would have the capacity for "perception."[19] If they exist, monads would count as spatial entities insofar as they could be the bearers of spatial relations. For example, the monads in the swarm that "compose" your pen would bear certain spatial relations to the monads in the swarm that "compose" your desk. The directionalist/relationalist theory of space (developed in chapter 3) and the reformulation of my answer to the Special Composition Question (developed in chapter 5) leave open the question of whether monads exist.

Souls. In addition to leaving open the question of whether there exist *part*-less entities that include mereological simples, zero-dimensional boundaries, and monads, I also leave open questions of whether *part*-less Cartesian *souls* exist and whether they would count as spatial entities. The directionalist/relationalist theory of space to be developed *does* allow that there exist dimensionless substances with the capacity for consciousness that bear spatial relations to other spatial entities.[20] For example, the theory allows that there exists a soul ½ meter to the right of the pen and ¼ meter to the left of the lamp. Souls would not be subject to the laws of physics as we know them; but no stand is taken here on whether this alone implies that souls would be thinking substances that are *nonphysical*.[21]

Events. Finally, the directionalist/relationalist theory of space leaves open the possibility that events exist and would count as spatial entities. For example, if, on Mount Katahdin, Susan finishes her four-month hike of the Appalachian Trail just as Robert begins his A.T. hike on Springer Mountain, then the directionalist/relationalist theory of space would allow that *Susan's completing the A.T.* is a concrete event that bears the spatial relation *occurs 3510 km to the northeast of* to *Robert's beginning the A.T.* As noted in the following chapter, I presuppose a particular theory of events for the sake of explicating the directionalist/relationalist theory of space.

NOTES

1. Def.8, Schol.II, *Sir Isaac Newton's Mathematical Principles of Natural Philosophy and his System of the World*, trans. Andrew Motte, rev. by Florian Cajoli (Berkeley: University of California Press, 1934), p. 6. See also George Berkeley, 111, *A Treatise Concerning the Principles of Human Knowledge*, ed. T. E. Jessop, Vol. II, *The Works of George Berkeley, Bishop of Cloyne*, ed. A. A. Luce and Jessop (London: Thomas Nelson and Sons, 1949): "[T]his celebrated author [Newton] holds there is an *absolute space*, which, being unperceivable to sense, remains in it self similar and immoveable. . . . *Place* he defines to be that part of space which is occupied by any body."

2. Thomas Reid, *Essays on the Intellectual Powers of Man*, ed. Ronald E. Beanblossom and Keith Lehrer (Indianapolis, IN: Hackett Publishing Co., 1983),

pp. 193–94. Cf. Thomas Hobbes, *Elements of Philosophy: Concerning the Body*, ed. Sir William Molesworth (London: John Bohn, 1839), II.VII.2, p. 93: "For no man calls it space for being already filled, but because it may be filled; nor does any man think bodies carry their places away with them, but that the same space contains sometimes one, sometimes another body." Though this passage suggests that Hobbes endorses substantivalist space, he soon after suggests a subjective account of space: "SPACE *is the phantasm of a thing existing without the mind simply*" (p. 94). See Gary B. Herbert, "Hobbes's Phenomenology of Space," *Journal of the History of Ideas* 48 (1987), 713: "Hobbes anticipates Kant's treatment of space as the outer form of sensuous intuition but without any notion of its existing *a priori* or in-itself."

3. Cf. Robin Le Poidevin, *Travels in Four Dimensions* (Oxford and New York: Oxford University Press, 2003), p. 37: "Space contains objects, rather like a box, only in this case we suppose the box to have no sides."

4. As formulated, SS leaves open the question of whether the whole of *space* is composed of a dense array of spatial points or whether *space* is "gunky" (i.e., whether every region of space is such that it has a subregion that itself has a subregion). Cf. Arntzenius, *Space, Time, and Stuff*, pp. 125–52.

5. See Peter Simons, "Where It's At: Modes of Occupation and Kinds of Occupant," *Mereology and Location*, ed. Shieva Kleinschmidt (Oxford and New York: Oxford University Press, 2014), pp. 59–60: noting that his "sympathies have always been on the ontologically sparser relationist side of this [substantivalist/relationalist] dispute," Simons confesses that "for the sake of simplicity and cognitive accessibility," his "forms of expression" are "*prima facie* substantivalist," and that this "does not entail that the talk should be taken at face value, but equally it does not entail that we should be able to eliminate such talk by paraphrase." In *Mereology and Location*, see also Daniel Nolan, "Balls and All," pp. 91–116, and see Peter Forrest, "Conflicting Intuitions About Space," pp. 117–31. Forrest develops an inconsistency involving our intuitions about the structure of "three-dimensional space," noting that his claims "may be adapted to spacetime of four or more dimensions". Finally, see Roberto Casati and Achille C. Varzi, *Parts and Places: The Structures of Spatial Representation* (Cambridge, MA, and London: MIT Press, 1999), pp. 3, 119. Though Casati and Varzi profess to "remain as neutral as possible as to 'real space' and its properties," they write as if committed to substantivalist space, claiming that "*your present temporary minimal address gives your exact location at this moment of time*, the region of space presently taken up by your body."

6. Richard Cartwright, "Scattered Objects," *Analysis and Metaphysics: Essays in Honor of R. M. Chisholm*, ed. Keith Lehrer (Dordrecht and Boston: D. Reidel Publishing Co., 1975), p. 153.

7. Ned Markosian, "A Spatial Approach to Mereology," *Mereology and Location*, pp. 70, 84.

8. Leibniz, *The Leibniz-Clarke Correspondence*, ed. H. G. Alexander (Manchester and New York: Manchester University Press, 1956), L.III.4 and L.V.47, pp. 25–26, 69.

9. Michael J. Loux and Dean W. Zimmerman, introduction to *The Oxford Handbook of Metaphysics* (Oxford and New York: Oxford University Press, 2003), p. 4: "Quine's criterion of ontological commitment is understood to be something like this: If one affirms a statement using a name or other singular term, or an initial phrase of 'existential quantification', like 'There are some so-and-sos', then one must either (1) admit that one is committed to the existence of things answering to the singular term or satisfying the description, or (2) provide a 'paraphrase' of the statement that eschews singular terms and quantification over so-and-sos."

10. For a succinct formulation of the difference between Spatial Substantivalism and Spatial Relationalism, see Shamik Dasgupta, "Substantivalism vs Relationalism About Space in Classical Physics," *Philosophy Compass* 10/9 (2015), 601–2; see also Tim Maudlin, "Buckets of Water and Waves of Space: Why Spacetime is Probably a Substance," *Philosophy of Science* 60 (1993), 184.

11. L.V.47, *The Leibniz-Clarke Correspondence*, pp. 70–71.

12. Hestevold, "Conjoining," pp. 371–85. As noted in the preface, I first raised the composition question and formulated Conjoining three years earlier in my 1978 dissertation.

13. See Franz Brentano, "On *Ens Rationis*," *Psychology from an Empirical Standpoint*, ed. Oscar Kraus, English ed. Linda L. McAlister (New York: Humanities Press, 1973), pp. 356–58. See also Brentano's *The Theory of Categories*, trans. Roderick M. Chisholm and Norbert Guterman (Boston: Martinus Nijhoff Publishers, 1981), p. 20 and Brentano's "On What is Continuous," *Space, Time and the Continuum*, trans. Barry Smith (London, New York, and Sydney: Croom Helm, 1988), pp. 1–44. See Roderick M. Chisholm, "Boundaries," *On Metaphysics* (Minneapolis: The University of Minnesota Press, 1989), pp. 83–89 (originally published as "Boundaries as Dependent Particulars" in *Grazer Philosophische Studien* 20 (1983)); and see Chisholm on "the problem of inner spatial contact" and "spatial dimensions" in *A Realistic Theory of Categories* (Cambridge: Cambridge University Press, 1996), pp. 87–91. See also Hestevold, "Boundaries, Surfaces, and Continuous Wholes," pp. 235–45.

14. David and Stephanie Lewis, "Holes," *Australasian Journal of Philosophy* 48 (1970), 206–12.

15. See Roberto Casati and Achille C. Varzi, *Holes and Other Superficialities*, A Bradford Book (Cambridge, MA, and London: The MIT Press, 1995).

16. I endorse the Doctrine of Unattached Parts. See Peter van Inwagen, "The Doctrine of Arbitrary Undetached Parts," *Pacific Philosophical Quarterly* 62 (1981), 123–37; William R. Carter, "In Defense of Undetached Parts," *Pacific Philosophical Quarterly* 64 (1983), 126–43. Cf. Dean W. Zimmerman, "Could Extended Objects Be Made Out of Simple Parts? An Argument for 'Atomless Gunk'," *Philosophy and Phenomenological Research* 56 (1996), 1–29; see also Ned Markosian, "Simples," *Australasian Journal of Philosophy* 76 (1998), 213–26 and "Simples, Stuff, and Simple People," *The Monist* 87 (2004), 405–42.

17. See Ned Markosian, "The Right Stuff," *Australasian Journal of Philosophy* 93 (2015), 665–87.

18. See Richard Cartwright's defense of scattered objects: "Scattered Objects," pp. 153–71. See also Chisholm's commentary on Cartwright's paper, "Scattered Objects," *On Metaphysics*, 90–95.

19. Sections 3,4, *Monadology*, ed. G. H. R. Parkinson, trans. Mary Morris and G. H. R. Parkinson (London: J. M. Dent & Sons Ltd, 1973), pp. 179, 180.

20. Cf. Joshua Hoffman and Gary Rosenkrantz, "Are Souls Unintelligible?" *Philosophical Perspectives*, Vol. 5, Philosophy of Religion (Atascadero, CA: Ridgeview Publishing Co., 1991), p. 197.

21. If souls *would* count as spatial entities, and if Ned Markosian is right that what distinguishes physical entities from nonphysical entities is that the former but not the latter *are* spatially located, then spatially located souls *would* be physical entities to which the laws of physics *do* apply. In this case, however, the (actual) laws of physics would be other than what we now believe them to be. (If *part*less souls *are* bearers of spatial relations and would thereby count as conscious *physical* entities, then the difference between a Cartesian soul and a Leibnizian monad would not be altogether clear.) See Ned Markosian, "What Are Physical Objects?" *Philosophy and Phenomenological Research* 61 (2000), 375–95.

Chapter 2

Spatial Directionalism

THE DIRECTIONED UNIVERSE

Leibniz argued that plausible claims that imply the existence of space, regions of space, places, or locations can be reformulated as claims that clearly imply nothing more than the existence of spatial entities and the spatial relations that obtain among them. Ontological economy is a sufficient reason to take Leibniz's view seriously. And, preservation of plausible claims about uniform movement, spatial orientation, and absolute motion is a sufficient reason to take seriously the Newton/Reid view that the world manifests some sort of "absolute spatial ordering." By countenancing the existence of certain spatial relations that Leibniz did not himself countenance—*spatial directional relations*—one can formulate a *directionalist* version of Spatial Relationalism that allows both the reduction of *space/place*-talk to *spatial-relations*-talk *and* the preservation of plausible claims about uniform movement, spatial orientation, and absolute motion. To explicate the concept of a spatial directional relation and to develop the directionalist/relationalist theory of space efficiently and consistently, I make several assumptions.

Relations. First, I assume that *relations* exist and are essentially *obtainable* abstract entities that exist necessarily. That is, I assume that every relation exists necessarily and is necessarily such that it is possible that there exist two entities between which the relation obtains. Thus, I would assume that *is a larger tree than* and *is a smaller unicorn than* are both relations that exist necessarily and that each relation is obtainable: it *is* possible that there do exist two trees such that one is larger than the other; and it *is* possible that there do exist two unicorns such that the first is smaller than the second. Presumably, of course, *is a larger tree than* is a relation that both exists *and* obtains whereas *is a smaller unicorn than* is a relation that exists but does not obtain.

States of affairs. Second, though I ultimately take no stand on the nature of events, I explicate the concept of a spatial directional relation and develop the directionalist/relationalist theory of space in terms of a states-of-affairs theory of events.[1] I assume that an event *is* a state of affairs that occurs where states of affairs, like relations, are abstract entities that necessarily exist. But not all existing states of affairs also *occur*: Whereas *someone thinking about space* both necessarily exists and contingently occurs, *an elephant chasing a unicorn* both necessarily exists and contingently does not occur. Presumably, the latter state of affairs has never occurred, nor will it ever occur unless unicorns come into being before elephants grow extinct. If, however, there is a possible world in which elephants *do* chase unicorns, then, in that world, *an elephant chasing a unicorn* would both necessarily exist *and* contingently occur. Presumably, *there existing at least one state of affairs* is itself a state of affairs that both necessarily exists and necessarily occurs. This states-of-affairs theory allows for *recurring* events. For example, *someone being inaugurated U.S. president* is likely not occurring as you read this sentence, but it first occurred in 1789 and has since recurred many times every four years; and it will occur again in January of the year that follows the next U.S. presidential election.[2] The directionalist/relationalist theory of space to be developed does not turn on this particular theory of events; I simply adopt this theory for the sake of writing efficiently and consistently about location vis-à-vis occurring events.

Spatial directional relations. One can informally explicate *Spatial Directionalism* (SD)—the view that the world is *spatially directioned*—in terms of relations, events, and objects: For any two events or objects that do not spatially overlap, one of the events/objects occurs/exists in a specific spatial direction relative to the other; and the other occurs/exists in the opposite spatial direction relative to the former. (SD is formulated more precisely at the end of the first section of chapter 3.)

Spatial Directionalism is on a par with Temporal Directionalism—the view that the world is *temporally* directioned. I assume that the world is temporally directioned in the sense that, for any two states of affairs that did, do, or will occur without temporally overlapping, one of the states of affairs did, does, or will occur in a specific temporal direction (e.g., *after*) relative to the other, and the other did, does, or will occur in the opposite temporal direction (e.g., *before*) relative to the former. The world, then, is *temporally* directioned in the sense that there exist temporal *directional* relations—for example, *is occurring after, did occur before, will occur before*—that obtain among existing states of affairs.[3] Similarly, then, to claim that the world is *spatially* directioned is to claim that there do exist spatial *directional* relations—for example, *is obtaining/existing direction D_{126} of* and *is obtaining/existing direction D_{483} of*—that obtain among existing states of affairs and objects.

Imagine a stark, spatially directed possible world in which two mereological simples, E_1 and E_2, are the only spatial entities that exist. In this world, exactly two spatial directional relations would obtain at any given time: The first entity would bear a specific directional relation (e.g., *is D_{126} of*) to the second, and the second would bear the opposite relation (i.e., *is D_{126*} of*) to the first.

How many spatial directional relations are there? Imagine two mereological simples such that one ceases to exist one minute before the second comes into being. There are exactly two ways that these two nontemporally overlapping mereological simples could be temporally separated by one minute (i.e., could be one minute apart): Either the first simple ceases to exist one minute *before* the second comes into being, or the second ceases to exist one minute *before* the first comes into being. With respect to spatial separation, however, there are indefinitely many ways that, at a given time, two discrete mereological simples could be spatially separated by one meter (i.e., could be one meter apart): The first could be one meter due north of the second or *vice versa*, the first could be one meter 2.0 degrees north-northeast of the second or *vice versa*, the first could be one meter 2.01 degrees north-northeast of the second or *vice versa, ad indefinitum*. One can state the intended implication of this example more exactly in terms of (absolute) spatial directional relations: In a directed universe, there exist indefinitely many spatial directional relations (independent of one's compass and one's location on Earth); and each spatial directional relation is such that, possibly, it is the only spatial directional relation that the first mereological simple bears to the second or that the second bears to the first.

Consider a second example that motivates the view that a directed universe would include indefinitely many spatial directional relations. Instead of imagining two mereological simples, imagine a one-meter sphere that is composed of indefinitely many densely packed mereological simples. That sphere's surface would itself include indefinitely many pairs of antipodal mereological simples one meter apart; and each of those indefinitely many simples would bear a unique spatial directional relation to its antipodal simple. So, there would thereby exist indefinitely many different spatial directional relations that would obtain among the simples that compose that sphere.

Of course, the spatial directional relations that obtain between any two of the sphere's antipodal mereological simples could change. Suppose that mereological simples E_1 and E_2 are two of the sphere's antipodal simples and that E_1 bears spatial directional relation *is D_{126} of* to E_2 and that E_2 bears opposite relation *is D_{126*} of* to E_1. Assume as well that mereological simple E_3

is located in the center of the sphere such that (a) E_1 bears *is* D_{126} *of* to E_3, and E_3 bears *is* D_{126} *of* to E_2 and (b) E_2 bears opposite spatial directional relation *is* D_{126*} *of* to E_3, and E_3 bears *is* D_{126*} *of* to E_1. Imagine now that the sphere rotates exactly 180° around any axis that is perpendicular to the E_1/E_2 axis. After the rotation, E_1 will no longer bear *is* D_{126} *of* to E_3 and to E_2, and E_2 will no longer bear *is* D_{126*} *of* to E_3 and to E_1. But E_2 *will* bear *is* D_{126} *of* to E_3 and to E_1, and E_1 *will* bear *is* D_{126*} *of* to E_3 and to E_2. If the sphere then rotates an additional 90°, then neither *is* D_{126} *of* nor *is* D_{126*} *of* will obtain among E_1, E_2 and E_3, but these two relations *will* obtain between two of the sphere's other antipodal simples.

How many spatial relations would obtain between two spatial entities? Although Spatial Directionalism implies that there exist indefinitely many spatial directional relations and that exactly two would obtain between any two mereological simples at a given time, Spatial Directionalism also implies that indefinitely many spatial directional relations would obtain between any two spatial entities when at least one of the two is non-zero-dimensional.

Assume that Mississippi (MS) and Alabama (AL) are two materially solid land masses and that there is a mereological simple M that is part of northern MS: With respect to any *non*-zero-dimensional part of northwestern AL, there would be indefinitely many specific spatial directional relations that M bears to that part alone. After all, *if* that non-three-dimensional AL part were composed of indefinitely many densely packed mereological simples, then M would bear a different spatial directional relation to each of those simples (and each of those simples would bear the opposite relation to M). For the same reason, (a) indefinitely many specific spatial directional relations would obtain between M and any non-zero-dimensional part of northeastern or central or southwestern AL, and (b) indefinitely many specific spatial directional relations would obtain between any *non*-zero-dimensional part of MS and any zero-dimensional or non-zero-dimensional part of AL.

Presumably, of course, there exist indefinitely many *different* spatial directional relations that do not obtain at all—specific spatial relations that *no* spatial entity bears to another. If, as astronomers tell us, the universe is expanding ever more rapidly, there are indefinitely many spatial directional relations that do not now obtain between any two spatial entities, but these relations *will* obtain as spatial entities continue to expand, moving farther and farther apart from one another.

DIRECTIONALISM, TIMES, AND PLACES

That the universe is temporally and spatially directed does not alone imply that there are or are not substantivalist *times* and *places*.

With respect to time, defenders of the A-Theory (i.e., Transient Time, Dynamic Time) defend "temporal passage" ("temporal becoming," "the flow of time").[4] These friends of "temporal passage" could all agree that the world is temporally directed while leaving open the question of whether substantivalist *times* exist. That is, A-Theorists could all agree that there *do* exist indefinitely many temporal *directional* relations (e.g., *is obtaining after*, *is obtaining before*) that can obtain among states of affairs while taking no stand on whether there exist *times* at which states of affairs would exist or occur and while taking no stand on whether *times* would themselves be bearers of temporal directional relations. For example, those A-Theorists sympathetic with Presentism could consistently endorse Temporal Substantivalism, arguing that "the privileged present" is the unique *time* (or sequence of *times*[5]) *at which* all present spatial entities do exist/occur. Such Substantivalist Presentists would claim that only the present *time(s)* exists, adding that there *did* exist and *will* exist *times* other than those that are present.[6] And, those A-Theorists who endorse both the Growing-Block Theory *and* Temporal Substantivalism would claim that past and present *times* all exist, adding that there *will* exist future times (that will then exist forevermore).[7]

Alternatively, those A-Theorists who find the concept of a substantivalist *time* problematic could consistently endorse the ontologically leaner Temporal Relationalism, arguing that all talk about the temporal location of objects and events can be reduced to talk about temporal directional relations.[8] For example, while insisting that no *time* exists, the Presentist/Relationalist could claim that *Socrates drinking hemlock* and *Socrates dying* both presently exist, that both states of affairs *did* occur, and the former presently bears the relation *occurred before* to the latter.

By the same token, those who reject "temporal passage" can consistently leave open the question of whether substantivalist *times* exist, siding with either Temporal Substantivalism or Temporal Relationalism. For example, while rejecting "temporal passage" and "the privileged present," such an Eternalist (i.e., B-Theorist, Static Time Theorist) could affirm both that the world is temporally directed and that substantivalist times exist. This Eternalist/Substantivalist could claim that there tenselessly/eternally exist *times* over which *Socrates drinking hemlock* tenselessly/eternally obtains and that each of these *times* (tenselessly/eternally) bears the relation *exists before* to the tenselessly/eternally existing times over which *Socrates dying* tenselessly/eternally obtains. On the other hand, an Eternalist who resists substantivalist *times* could consistently endorse Temporal Relationalism, claiming that spatial entities, states of affairs, and temporal directional relations all exist tenselessly and eternally and that all *time*-talk can be reduced to talk involving nothing more than temporal directional relations and those spatial entities among which such relations obtain. Without reference to substantivalist *times*, such an Eternalist/Relationalists could claim that the temporal directional relation *tenselessly ceases to exist before the coming into being of* obtains tenselessly between two existing buildings, one of which is tenselessly demolished on the same piece of land on which the other is tenselessly constructed. And, Eternalist/Relationalist could claim as well that the temporal directional relation *occurs after* obtains tenselessly and eternally between *the eruption of Shira's first baby tooth* and *Shira's birth*.

Just as commitment to a temporally directed world is consistent with both the existence and the nonexistence of substantivalist *times*, commitment to a *spatially* directed world is consistent with both the existence and the nonexistence of substantivalist *places*—substantivalist regions of *space*. Assume that Star X comes into being simultaneously with the collapsing of Star Y, and assume that there exist spatial directional relations that include the specific relations *exists direction d* of* and *obtains direction d* of*. One can consistently endorse Spatial Directionalism and Spatial Substantivalism, claiming that Star X comes into being in a region of substantivalist *space* that bears the spatial directional relation *exists direction d* of* to the region of substantivalist *space* where Star Y collapses. Alternatively, the Spatial Directionalist could side with Spatial Relationalism and report instead that *Star X's coming into being* and *Star Y's collapsing* are states of affairs that occur simultaneously and that the former bears the spatial directional relation *obtains direction d* of* to the latter. That a spatial directional relation obtains between two spatial entities would not *ipso facto* imply that there do or do not exist substantivalist regions of space *where* those spatial entities exist or occur. Left open is the question of whether regions of *space* exist and are among those entities among which spatial directional relations can obtain.

SPATIAL DIRECTIONAL RELATIONS

Though the concept of a spatial entity's bearing a certain spatial directional relation to another spatial entity is taken as primitive, I will explicate the concept informally (referring to figure 2.1) and will then offer several axioms to distinguish spatial directional relations from other relations.

In figure 2.1, let A, B, and C represent three zero-dimensional mereological simples; and let D, E, and F represent three nonoverlapping three-dimensional materially solid cubes. For the sake of explicating the concept of one spatial entity's bearing a spatial directional relation to another, assume first that simples A and B compose scattered object AB, that B and C compose scattered object BC, and that AB and BC are overlapping parts of scattered whole ABC. Second, assume also that materially solid spatial entity DEF is located several centimeters away from ABC and that D, E, and F are

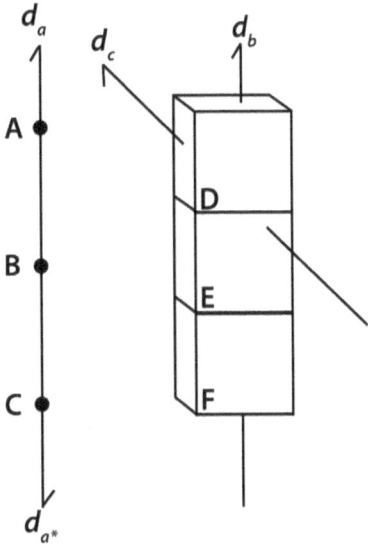

Figure 2.1 Spatial Directional Relations. *Source*: C. Frantom, J. Rich: White Roche LLC.

*non*overlapping proper parts of DEF, whereas DE and EF are proper parts of DEF that *do* overlap.

Where d_a represents the only spatial directional relation that mereological simple A bears to B, that B bears to C, and that A bears to scattered spatial entity BC, d_{a*} represents the opposite spatial directional relation, which is the only such relation that C bears to B, that B bears to A, and that C bears to scattered spatial entity AB. Where d_b is a spatial directional relation other than d_a such that D bears d_b to E and to F and to EF, there also exist indefinitely many other spatial directional relations that D bears to E. And, every relation that D bears to E is also a relation that D bears to EF. There are, however, relations (e.g., d_c) that D bears to E and to EF that neither D nor E bears to F.

Similarly, mereological simple B bears indefinitely many spatial directional relations to E, and each of the relations that B bears to E is a relation that B also bears to DE, to EF, and to DEF. Of course, B bears indefinitely many spatial directional relations to D that it does not bear to certain other proper parts of DEF. For example, B bears a spatial directional relation to the bottom half of F; and although this is a relation that B also bears to EF and to DEF, it is not a relation that B bears to E, to D, or to DE.

A necessary condition for one entity's bearing a spatial directional relation to another is that, with respect to that relation, the two are "entirely separated" in the sense that the former "in its entirety" bears the relation to the latter. Put another way, the two are "entirely apart" such that no parts of the former "surround" parts of the latter. To develop the sense in which one spatial entity "in its entirety" bears a spatial directional relation to another, consider the sense in which one may report accurately that Wyoming is "in its entirety" to the north of Colorado, whereas Nebraska is *not* "in its entirety" to the east of Colorado. Loosely, Wyoming is "completely north" of Colorado in the sense that no part of Wyoming "falls below" the northern border of Colorado—no part of Wyoming "surrounds" the western or eastern border of Colorado and no protruding part of Wyoming is surrounded by a part of Colorado. By the same token, Nebraska is "entirely" east of Wyoming in the sense that no part of Wyoming "extends over" the western border of Nebraska. And, Nebraska is *not* "entirely east" of Colorado: Part of Nebraska "surrounds" the northwest corner of Colorado.

To understand more clearly the sense in which one spatial entity bears "in its entirety" a certain spatial directional relation to another, refer to figure 2.2. Letting G-Q represent nonoverlapping three-dimensional materially solid wholes, assume that materially solid spatial entity GHIJKLM is a T-shaped spatial entity composed of GHIJK and LM and that GHIJKLM's proper parts include scattered spatial entities GLK, GL, GLM, LK, MLK, and GILMK. Finally, assume also that N and O compose NO, that P and Q compose PQ,

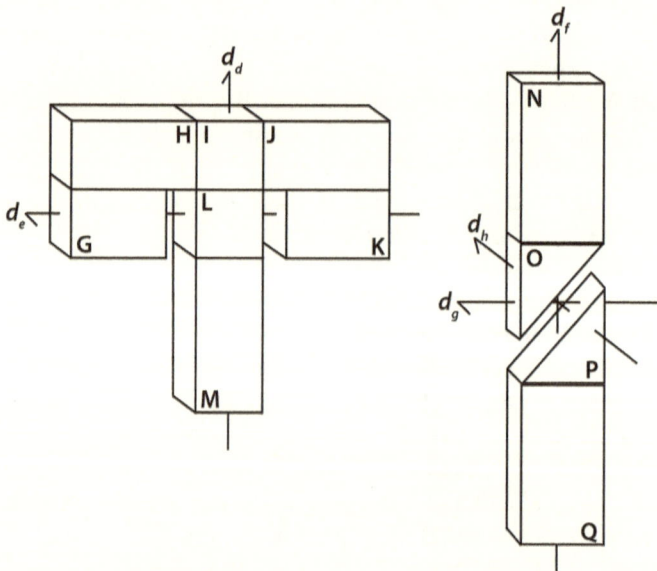

Figure 2.2 Spatial Entities Bearing Spatial Directional Relations to Other Spatial Entities. *Source*: C. Frantom, J. Rich: White Roche LLC.

that O and P compose scattered spatial entity OP, and that NO and PQ compose scattered spatial entity MNOP.

As is the case with DEF in figure 2.1, spatial entity I in figure 2.2 bears d_d to L and to LM; and IL and L both bear d_d to M. But GHIJK does *not* bear d_d to L or to LM: Just as Wyoming is not "in its entirety" north of Utah, GHIJK does not "in its entirety" bear d_d to L or to LM given that parts of GHIJK "surround" at least *some* proper parts of L and LM. Put more technically, with respect to d_d, there is at least one perpendicular spatial directional relation (e.g., d_e) such that a proper part of GHIJK bears that relation to a proper part of LM. When two spatial entities *are* "entirely separated" with respect to a given spatial directional relation, neither has a proper part that bears a perpendicular spatial directional relation to a proper part of the other. Given that d_d is perpendicular to d_e and that I is a proper part of GHIJK that bears d_d to indefinitely many proper parts of L, it is false that GHIJK bears d_e to L— GHIJK does not "in its entirety" bear d_e to L. G does, however, bear "in its entirety" d_e to L, to IL, to LM, to ILM, to LK, to ILK, to LMK, and to ILMK.

With respect to scattered spatial entity NOPQ in figure 2.2, N "in its entirety" bears d_f to P, to PQ, and to Q. And, NO and O each bears "in its entirety" d_f to Q but not to P or to PQ: With respect to d_f, there is at least one perpendicular relation (e.g., d_g) such that proper parts of NO and O bear the relation to proper parts of P and PQ. That is, with respect to spatial directional relation d_f, NO and O are not "entirely separated" from P and PQ; rather, NO and O partially "surround" parts of P and PQ. Spatial entities O and NO do, however, bear "in their entirety" d_h to P and to PQ: With respect to d_h, there is no perpendicular spatial directional relation that a proper part of O or NO bears to a proper part of P or PQ.

With an understanding of when it is that a spatial directional relation does and does not obtain between two "entirely separated" spatial entities, consider several axioms that distinguish spatial directional relations from other relations.

Spatial directional relations are transitive:

SDR$_1$ For any spatial directional relation d_n and time t, if x bears d_n to y at t, and *if* y bears d_n to z at t, then x bears d_n to z at t.

With respect to figures 2.1 and 2.2, the above axiom of transitivity implies that D bears d_b to F and that N bears d_f to Q given that, respectively, (i) D bears d_b to E, and E bears d_b to F, and (ii) N bears d_f to O, and O bears d_f to Q. (Although N bears d_f to O, SDR$_1$ does *not* imply that N bears d_f to P given that O does not bear d_f to P: O and P are not "entirely separated" with respect to d_f.)

Spatial directional relations are also asymmetric: With respect to any spatial directional relation that one spatial entity bears to a second, the second does *not* bear that relation to the first:

SDR$_2$ For any spatial directional relation d_n and time t, if x bears d_n to y at t, then it is false that y bears d_n to x at t.

Regarding figures 2.1 and 2.2, there are indefinitely many spatial directional relations that, respectively, B bears to E and that I bears to M, but neither E nor M bears any one of those relations to B or I. Of course, for any one spatial directional relation that B bears to E or that I bears to M, there would be a single "exact opposite" spatial directional relation that E would bear to B or that M would bear to I.

Spatial directional relations are not only transitive and asymmetric but irreflexive:

SDR$_3$ For any spatial directional relation d_n and time t, it is false that there exists something x such that x bears d_n to x at t.

With respect to figure 2.1, there are indefinitely many spatial directional relations that D bears to E, to F, and to EF, but there is *no* spatial directional relation that D bears to DE given that D is not "entirely separated" from DE and thereby cannot "in its entirety" bear any spatial directional relation to DE. Similarly, no spatial entity can be "entirely separated" from itself, and thereby no spatial entity can bear to itself a spatial directional relation that it may well bear to indefinitely many other spatial entities.

One may be tempted to reject SDR$_3$, insisting that spatial directional relations *can* be reflexive: "Spatial directional relations are *not* essentially irreflexive. Suppose that figure 2.3 represents a materially solid sphere Q composed of an inner spherical core R surrounded by a nonoverlapping outer shell S. Consider spatial directional relation d_i, which is among the indefinitely many spatial directional relations that the sphere's bottom hemisphere bears to its top hemisphere. If Q's bottom hemisphere bears d_i to the top hemisphere, then (in virtue of S's having proper parts "below" R's bottom-half) S would itself bear d_i to R's bottom-half; and R's bottom-half would bear d_i to R's top-half, and R's top-half would bear d_i to S (in virtue of S's having proper parts "above" R's top-half). If, however, S bears d_i to R's bottom-half, R's bottom-half bears d_i to R's top-half, and R's top-half bears d_i to S, then the axiom of transitivity would imply that S bears d_i to S. If S *would* bear d_i to S with respect to sphere Q composed of S and R, then spatial directional relation d_i is *reflexive* with respect to sphere Q; and SDR$_3$ should thereby be rejected."

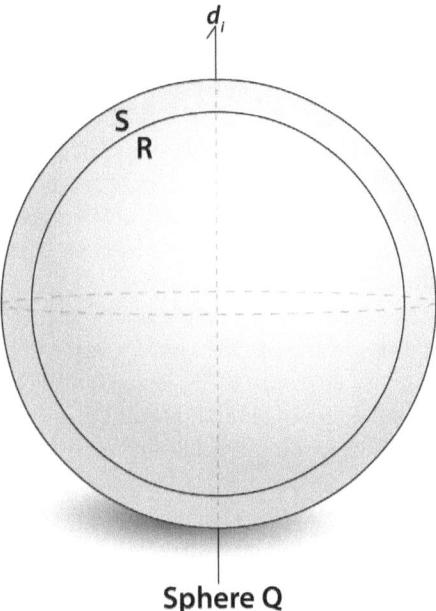

Figure 2.3 The Irreflexivity of Spatial Directional Relations. *Source*: C. Frantom, J. Rich: White Roche LLC.

This objection to SDR_3 rests on a mistake: It is false that S bears d_i to R's bottom-half, and it is false that R's top-half bears d_i to S. Spatial entity S "completely surrounds" R's bottom-half and is thereby not "entirely separated" from R's bottom-half; so, it is false that S "in its entirety" bears d_i to R's bottom-half. (There is, of course, a relatively small proper part of S located "in its entirety" below R, and that proper part bears d_i to R's bottom-half, to R's top-half, and to S's top-half.) It is similarly false that R's top-half bears d_i to S: Because S "completely surrounds" R, Q's top-half is not "entirely separated" from S and thereby does not bear d_i to S: R's bottom-half does *not* "in its entirety" bear d_i to S. Thus, the axiom of symmetry does not imply that S bears d_i to S given that S does *not* bear d_i to R's bottom-half and R's top-half does not bear d_i to S.

Spatial directional relations are such that, if one obtains, then there exists a second—"the opposite direction"—that obtains, if and only if, the first obtains:

SDR_4 For any spatial directional relation d_n, there exists exactly one other spatial directional relation d_{n*} such that, for any x and y at any time t, x bears d_n to y at t, if and only if, y bears d_{n*} to x at t.

With respect to any two three-dimensional materially solid spatial entities, there are indefinitely many spatial directional relations that each bears to the other: One entity's leftmost eighth bears multiple spatial directional relations to the other's rightmost eighth, and the leftmost eighth bears still other such relations to the other entity's uppermost eighth. Yet, for any one of these spatial directional relations, there is a unique ("opposite" or "reciprocal") spatial directional relation such that the one obtains between two spatial entities, if and only if, the other also obtains between those two entities. Similarly, with respect to mereological simple B and materially solid DEF represented in figure 2.1, the former bears indefinitely many spatial directional relations to D, to E, to F, and to other proper parts of DEF; but for any one of those relations, B bears that relation to a proper part of DEF, if and only if, there exists a certain other ("opposite") spatial directional relation that DEF's proper part bears to B.

Consider a fifth axiom regarding spatial directional relations:

SDR$_5$ For any spatial directional relation d_n, if, at any time t, (i) x bears d_n to y, (ii) x does not bear d_n to z, and (iii) z does not bear d_n to x, then, at t, y does not bear d_n to z and z does not bear d_n to y.

SDR$_5$ implies that if Tennessee bears a certain spatial directional relation to Alabama and that relation does *not* obtain between Tennessee and North Carolina, then that relation does not obtain between Alabama and North Carolina. By the same token, if there is a certain spatial directional relation that obtains between Alabama and Kentucky—and there *is*—then either Tennessee does *not* bear that relation to Alabama or that relation does obtain between Tennessee and Alabama. Indeed, there *are* indefinitely many spatial directional relations that Kentucky bears to Alabama; and many of these are also relations that obtain between Tennessee and Alabama.

SPATIAL ENTITIES

Hobbes claimed that a physical object is an entity that, *"having no dependance upon our thought, is coincident or coextended with some part of space."*[9] To sidestep ontological commitment to regions of space, I suggest instead that a necessary condition of an entity's counting as a physical object is that it could be the bearer of spatial directional relations—that it is possible that it bears at least one spatial directional relation to some other entity. Taking no further stand on what physical objects' necessary and sufficient conditions may include and leaving open the possibility that there exist *non*-physical objects that are the bearers of spatial directional relations, I explicate the concept of a *spatial entity* in terms of the primitive concept of one entity's bearing a spatial directional relation to another:

D2.1 x is a spatial entity =Df It is logically possible that there exists something to which x bears a spatial directional relation.

Spatial entities, then, are entities that *could* be bearers of spatial directional relations. If *souls* exist and if a soul is closer "in a certain direction" to a person's pancreas than it is to the Moon "in some other direction," then this nonphysical "thinking substance" would be the bearer of spatial directional relations and would thereby count as a nonphysical spatial entity. Similarly, if Leibniz's (dimensionless) monads exist, then they would also count as spatial entities: For any two monads, each would bear a unique spatial directional relation to the other at any given time.

A spatial directional relation *obtains* between two spatial entities when one of the two entities bears that relation to the other:

D2.2   Spatial directional relation d_n obtains between x and y at time t =Df At time t, where d_n is a spatial directional relation, either x bears d_n to y or y bears d_n to x.

If a three-dimensional materially solid cube exists, then it would be a spatial entity such that indefinitely many spatial directional relations would obtain "internally" among its proper parts: indefinitely many spatial directional relations would obtain between its top-half and bottom-half, indefinitely many others would obtain between its left- and right-halves, and still indefinitely many others would obtain between its uppermost rear eighth and its lowermost front eighth.

If there exists a zero-dimensional mereological simple, then *no* spatial directional relation would obtain "internally" with respect to that simple: Such a simple would have *no* parts at all between which spatial directional relations could obtain. Moreover, if a given mereological simple is the *only* spatial entity that exists, then *no* spatial directional relation would obtain at all: Not only would the simple lack proper parts between which such relations could obtain "internally," but there would exist no other spatial entity to which that solitary simple would bear such relations "externally." In such a stark world, the lone simple would nonetheless count as a spatial entity: It is at least logically possible that there exists something to which the simple bears a spatial directional relation.

In the following chapter, as I formulate Spatial Directionalism and develop a directionalist/relationalist theory of space, I leave open the questions of whether souls, monads, and mereological simples exist, and I leave open the question of whether there is a stark possible world in which the only spatial entity is a mereological simple.

NOTES

1. See, for example, Roderick M. Chisholm, *Person and Object: A Metaphysical Study* (LaSalle, IL: Open Court Publishing Co., 1976), pp. 114–31. Cf. Chisholm's reduction of states of affairs to properties, "Properties and States of Affairs," *On Metaphysics*, pp. 148–49; and, in the same volume, see his theory of events cast in terms of attributes and individuals, "States and Events," pp. 150–55.

2. See Major L. Johnson, Jr., "Events as Recurrables," *Analysis and Metaphysics: Essays in Honor of R. M. Chisholm*, pp. 209–26; and see Chisholm, "Events and Propositions," *Noûs* IV (1970), 15–24. Cf. Lawrence Lombard, *Events: A Metaphysical Study* (London, Boston, and Henley: Routledge & Kegan Paul, 1986), p. 131.

3. For an overview of temporal relations vis-à-vis the dispute between temporal substantivalists and relationalists, see L. Nathan Oaklander, *The Ontology of Time* (Amherst, NY: Prometheus Books, 2004), pp. 20–26, 193–203. See also Craig Bourne, *A Future for Presentism* (Oxford: Clarendon Press, an imprint of Oxford University Press, 2006), pp. 95–99. My assumption that the world is temporally directed—the assumption that, as Huw Price frames it, "it is an objective matter which of two nonsimultaneous events is the earlier and which the later"—is controversial. Price explains his own concerns about this view in "The Flow of Time," *The Oxford Handbook of Philosophy of Time*, ed. Craig Callender (Oxford: Oxford University Press, 2011), pp. 277, 281–302, 341–43. In the same volume, resisting the view that "the fixity of the past [is] a strict, fundamental, metaphysical or scientific fact" (p. 247), Douglas Kutach appeals to agency regarding the advancing of one's goals to explain the apparent difference between the earlier and the later; see "The Asymmetry of Influence," pp. 247–75. In the *The Oxford Handbook of Philosophy of Time*, Jill North claims that one can appeal to what one takes to be "the best account of the thermodynamic asymmetry" to infer that time does or does not have an objective inherent structure—that there *is* or is *not* a fundamental difference between earlier and later events; see "Time in Thermodynamics," pp. 312–50. See Paul Horwich on the anisotropy of time without "*privileged* direction"; *Assymetries in Time: Problems in the Philosophy of Science*, A Bradford Book (Cambridge, MA, and London: The MIT Press, 1987), pp. 37–45. See also Arntzenius, *Space, Time, and Stuff*, pp. 19–38. Cf. Dean W. Zimmerman, "The Privileged Present: Defending an 'A-theory' of Time," *Contemporary Debates in Metaphysics*, ed. Theodore Sider, John Hawthorne, and Zimmerman (Malden, MA: Blackwell, 2007), pp. 221–22. To defend the view that time is directed—that there is a fundamental difference among the past, present, and future such that past happenings are *before*, and future happenings are *after*, present happenings—Zimmerman would appeal to Presentism, which is the view that anything that exists/occurs *presently* exists/occurs. Claiming

that Presentism "is simply *part of commonsense* that the past and future are less real than the present," Zimmerman notes that those who deny this commonsense view constitute "a relatively small group of people—basically, people who have become accustomed to using spatial metaphors to understand temporal notions (as one does when drawing space–time diagrams or reading the more consistent science fiction stories about time travel)."

4. Defenders of Transient Time would include Presentists ("Only the present is real"), "growing-block" theorists ("Only the past and present are real, and only the present is *now*"), and "moving-spotlight" theorists ("The past, present, and future are real, only the present is *now*, and 'the now' moves from the present to the future and away from the past").

5. Whether "the present" has temporal extension is controversial. Those sympathetic with the specious present will allow that more than one time presently exists; others have argued that Presentists should agree that "the present" is not instantaneously, durationlessly thin, allowing for a "thick" present. I press the case for a "thick" present in "Presentism: Through Thick and Thin," pp. 330–31.

6. For example, such Presentists could claim that there *did* exist (but no longer exists) a time at which my birth occurred and that there *will* exist a time at which my death occurs.

7. For example, the Substantivalist Growing-Block Theorists would claim that there exist the *times* over which my birth occurred *and* the present *time*. They would also claim that there *will* exist a time at which my death occurs and that that *time* will become an existing past *time* after it ceases to be present.

8. When I defended the view that Presentists should endorse "an extended present," I also argued that Presentists should resist substantivalist *times*; and I then formulated Time-Free Thick Presentism in terms of temporal relations without reference to substantivalist times. See "Presentism: Through Thick and Thin," pp. 326–30.

9. Hobbes, II.8.1, *Concerning the Body*, p. 102. Cf. George Edward Moore, *Some Main Problems of Philosophy* (London: George Allen & Unwin; and New York: Humanities Press, 1956), p. 128: "[N]othing can be a material object except what has position in space."

Chapter 3

A Directionalist Theory of Space

DIRECTIONALIST LOCATIONS

Leibniz claimed that the concept of an entity's *place* can be analyzed in terms of the spatial relations that *that* entity bears to other entities:

> [T]o have an idea of place, and consequently of space, it is sufficient to consider these relations . . . without needing to fancy any absolute reality out of the things whose situation we consider. And, to give a kind of a definition: *place* is that, which we say is the same to A and, to B, when the relation of the co-existence of B, with C, E, F, G, etc. agrees perfectly with the relation of the co-existence, which A had with the same C, E, F, G, etc.[1]

In this chapter, I take seriously Leibniz's claim that an entity's place *is* the spatial relations that it bears to other entities. But unlike Leibniz, I assume that the world is spatially *directioned* and that the spatial relations that any entity bears to others would include spatial *directional* relations. According to this directionalist/relationalist view, then, a spatial entity's *place* at a given time is those spatial *directional* relations that obtain or could obtain between that entity and others at that time.

Richard Cartwright has used the term "receptacle" to refer to "a region of space with which it is possible some material object should be, in Hobbes' phrase, coincident or coextended."[2] Instead of countenancing the existence of *occupiable* regions of space that serve as the "receptacles" for material objects, consider positing instead *bear*-able sets of spatial directional relations that are the *possible locations* for material objects and for any other spatial entities that exist. Loosely, a possible location is a set of spatial directional relations that is *bear*-able with respect to a certain spatial entity. And, a set of relations is *bear*-able with respect to a certain entity if it is possible that the entity bears one of those relations to something while also bearing any one of the set's other relations to something else. Below is a more precise account of the concept of a *possible location*:

> D3.1 p is a possible location for spatial entity x =Df p is a set of all and only those spatial directional relations such that, for any member of p, d_n, it is logically possible both (i) that x bears d_n to something y at a time t and (ii) that, at time t, x bears any other member of p to something other than y.

Before appealing to the concept of a possible location to analyze the concept of an entity's *actual* location, a few words about sets are in order. For the sake of offering analyses that are more direct and more easily understood, I formulated D3.1 and will formulate others in terms of *sets*. Friends of realism, however, could reformulate D3.1 (and later *set*-involving definitions) in terms of *properties*:

> p is a possible location for spatial entity x =Df p is a property that is exemplified by all and only those spatial directional relations such that, for any spatial directional relation d_n that exemplifies p, it is logically possible both (i) that x bears d_n to something y at t and also that (ii) at t, x bears to something other than y any spatial directional relation that exemplifies p and is other than d_n.

Having noted that realists may insist on reformulating my *set*-involving analyses as *property*-involving analyses, I leave open the question of whether sets exist and turn to the concept of a spatial entity's actual location.

Though there would be many *possible* locations for any given spatial entity at a given time, there would be at that time but a single set of spatial directional relations that would be that spatial entity's *actual spatial location*. That is, for any spatial entity at a given time, there would exist only one possible location for that entity that would include those spatial directional relations that the entity would actually bear to other spatial entities:

> D3.2 p is the actual spatial location of x at time t =Df (i) p is a possible location for x, and (ii) at t, x and p are such that, for any y and for any spatial directional relation d_n, if x bears d_n to y at t, then d_n is a member of p.

D3.1 and D3.2 leave open the possibility that mereological simples and zero-dimensional boundaries exist and have actual locations.[3] For example, consider again the spatial entities represented in figure 2.1 and assume that they all exist at present time T. At T, there would exist indefinitely many possible locations for B: Each of the indefinitely many sets of spatial directional relations would be such that it is logically possible both that B bears a particular member of that set to a second spatial entity at T and that, at T, B also bears any other member of that set to yet a third spatial entity. B's *actual* spatial location at T would be the one possible location for B that includes the direction that B bears to C at T (i.e., d_a), the direction that B bears to A at T (i.e., d_{a*}), and every spatial directional relation that B bears to DEF and to any of DEF's parts. Given that A, B, and C are all mereological simples, D3.1 and D3.2 allow that the sets of spatial directional relations that are the actual spatial locations for A and C at T would also be among B's possible locations; and, at T, B's actual spatial location would also be a possible location for A and for C.[4]

To say of a spatial entity that it *exhibits externally* a particular spatial directional relation is to say that that relation obtains between that spatial entity and some other:

> D3.3 x exhibits externally spatial directional relation d_n at time t =Df At time t, there exists something y such that spatial directional relation d_n obtains between x and y.

Consider again figure 2.1 and suppose both that DEF is a materially solid object composed of three stacked cubes and that DEF and its parts are the only spatial entities that exist at a time t. Per D3.3, E would *exhibit externally* indefinitely many spatial directional relations at t, including those that obtain between E and D, between E and F, between E and D's upper-half, and between E and F's left third. If at t no spatial entity other than DEF and its parts exist, then DEF would itself exhibit *no* spatial directional relations externally: There would exist nothing to which DEF could bear such relations and nothing that would bear such relations to DEF.

Though DEF would itself exhibit externally *no* spatial directional relations at t (when DEF and its parts are the only spatial entities that exist), DEF *would* then exhibit *internally* indefinitely many spatial directional relations: There exist indefinitely many spatial directional relations that, at t, *would* obtain among nonoverlapping proper parts of DEF. The mereological concepts essential to the analysis of the concept of exhibiting spatial directional relations *internally* are themselves analyzable in terms of actual spatial locations:[5]

> D3.4 At time t, x's actual spatial location lies within w's actual spatial location =Df The actual spatial location of x at t is a proper subset of the actual location of w at t.

D3.5 *x* is a proper part of *w* at time *t* =Df *x* and *w* are necessarily such that, for any time *t* at which *x* and *w* exist, *x*'s actual spatial location lies within *w*'s actual spatial location.

Per D3.4 and D3.5, the set of spatial directional relations that *is* DEF's actual location at a given time would include all of those spatial directional relations that are members of the set of those spatial directional relations that *is* E's actual location at that time. And, as they should, D3.4 and D3.5 also allow that the proper part relation is transitive, asymmetric, and irreflexive: If D is a proper part of DE, and if DE is a proper part of DEF, then D is a proper part of DEF; but neither DE nor DEF is a proper part of D, and D is *not* a proper part of itself.

An important implication of D3.4 and D3.5 is that the concept of a proper part—a fundamental mereological concept—is reducible to concepts that involve sets of spatial directional relations: The concept of a proper part is analyzable in terms of actual spatial locations, the concept of an actual spatial location is analyzable in terms of the concept of a possible location, and the concept of a possible location is analyzable in terms of the concept of a set of spatial directional relations. I will address the significance of this reductive account of proper parts in chapter 5 when I defend my reformulated answer to the Special Composition Question.

Given that the concept of nonoverlapping (i.e., discrete) spatial entities can be analyzed in terms of the concept of a proper part, talk of nonoverlapping spatial entities is similarly reducible to talk of sets of spatial directional relations:

D3.6 *x* and *y* are nonoverlapping [discrete] spatial entities =Df (i) *x* and *y* are spatial entities such that *x* is other than *y*; (ii) there is no *z* that is a proper part of both *x* and *y*, and (iii) *x* is not a proper part of *y*, and *y* is not a proper part of *x*.[6]

Finally, the concept of a spatial entity's exhibiting a spatial directional relation internally can be analyzed in terms of nonoverlapping proper parts:

D3.7 Spatial entity *x* exhibits internally spatial directional relation d_n at time *t* =Df At time *t*, there exist two nonoverlapping proper parts of *x* such that d_n obtains between those two proper parts.

DEF would exhibit no spatial directional relations *externally* in the materially lean world in which DEF and its parts are the only spatial entities, but DEF *would*, in that world, exhibit indefinitely many spatial directional relations *internally* given that DEF would have nonoverlapping proper parts among which such relations would obtain. Worth noting is that D3.7 does leave open the possibility that mereological simples exist and that a materially solid spatial entity exhibits internally a spatial directional relation that obtains between two proper parts of which at least one is a dimensionless mereological simple. No mereological simple or any other *part*-less spatial entity can exhibit spatial directional relations internally: Such dimensionless spatial entities would have no proper parts among which spatial directional relations could obtain.

As one would expect, the concept of composition is another concept that is analyzable in terms of nonoverlapping proper parts:

D3.8 w is composed [i.e., strictly made up] of x and y =Df (i) x is a proper part of w, (ii) y is a proper part of w, (iii) x and y are nonoverlapping spatial entities, and (iv) no proper part of w fails to overlap x or y.

Consider again the various spatial entities represented in figure 2.1. D10 allows for the possibility that there does exist a spatial entity ABC that is composed of A and BC, that BC is composed of B and C, that DEF is composed of DE and F, that DE is composed of D and E, and that ABCDEF is composed of ABC and DEF. Though I endorse D3.8 as the analysis of what it would mean to say of a spatial entity that it is composed of two others, I will not address until chapter 5 the Special Composition Question—the question of under what conditions two spatial entities *do* compose (or *fail* to compose) a whole.

Earlier (in chapter 1), I loosely characterized Spatial Directionalism as the view that the universe is spatially directed—that there exist spatial directional relations that would obtain among any existent spatial entities. To formulate Spatial Directionalism more precisely, one should *not* characterize it as the view that every spatial entity exhibits at least one spatial directional relation externally or internally. After all, the possibility should be left open that, in a stark world in which a single mereological simple is the only spatial entity that exists, there would exist no other entity to which that mereological simple would bear any spatial directional relation, and that entity would have no proper parts between which any spatial directional relation could obtain. To formulate Special Directionalism more precisely, then, one should characterize it as the view that any spatial entity would *involve* at least one spatial directional relation:

D3.9 x involves spatial directional relation d_n at time t =Df There is a set p that is the actual spatial location of x at time t, and d_n is a member of p.

SD The world is *spatially directed*: There exist spatial directional relations; and any spatial entity that exists involves at least one spatial directional relation at any time that it exists.

DIRECTIONALIST REDUCTIONS

Leibniz claimed that, to skirt ontological commitment to substantivalist *space*, talk about *space* and talk about one entity's occupying the same *place* that another entity occupied earlier can be reformulated as talk about spatial relations:

> And supposing . . . that among those co-existents, there is a sufficient number of them, which have undergone no change; then we may say, that those which have such a relation to those fixed existents, as others had to them before, have now the *same place* which those others had. And that which comprehends all those places, is called *space*. Which shows, that in order to have an idea of place, and consequently of space, it is sufficient to consider these relations . . . without needing to fancy any absolute reality out of the things whose situation we consider. . . . Lastly, *space* is that, which results from places taken together.[7]

Though Spatial Directionalists would add that spatial *directional* relations are among the spatial relations that exist, a Spatial Directionalist could otherwise agree with Leibniz that claims about "places" and "empty space" are reducible to claims about the spatial (directional) relations that obtain or could obtain among spatial entities. Consider several paradigm examples of such reductions.

One may observe that a single place can be occupied by different spatial entities at different times:

C_1 Blair's car now occupies the same place that Rowan's car occupied yesterday.

By invoking D3.2—the directionalist analysis of an *actual spatial location*—one can avoid C_1's implication that there exists a *place* that has been occupied by two different cars:

$C_{1'}$ The present actual spatial location of Blair's car was the actual spatial location of Rowan's car yesterday: The set of spatial directional relations that, yesterday, was the actual spatial location of Rowan's car is now the actual spatial location of Blair's car.

In day-to-day discourse, one may also endorse reasonable claims about empty spaces and about the sizes of certain regions of space:

C_2 There is now a region of empty space as big as the Moon.

By invoking D3.1 and D3.2—the directionalist concepts of, respectively, *possible location* and *actual spatial location*—one may reduce C_2 to the following claim that implies the existence of nothing more than spatial entities, spatial directional relations, and sets of spatial directionalist relations:

$C_{2'}$ There exists now a set of spatial directional relations p such that (i) p is a possible location of the Moon, (ii) it is false that p is now the actual spatial location of any spatial entity, and (iii) for any subset p' of p such that it is possible that p' is the actual spatial location of a spatial entity, it is false that p' is the actual spatial location of any spatial entity.

The reductions above suggest that one may similarly reduce other *place-* and *space*-implying claims to relationalist claims that do not themselves imply the existence of substantivalist *space*.

THE DIRECTIONALIST THEORY OF SPACE

Those sympathetic with Spatial Relationalism have cause for optimism: If reasonable *place-* and *space*-implying claims *are* reducible to claims involving spatial directional *relations*, and if commitment to substantivalist *space* is itself a problematic, then endorsing a relationalist theory of space formulated in terms of spatial directional relations would seem reasonable. Consider, then, the Directionalist Theory of Space:

DTS It is false that *space* or regions of *space* exist; and spatial directional relations do exist such that reasonable claims that imply that *space* or regions of *space* exist can be reformulated as claims (involving spatial entities and spatial directional relations) that clearly do not imply that *space* or regions of *space* exist.

Assuming that all spatial *directionalist* relations *are* members of the set of all spatial relations, DTS implies, but is not implied by, Spatial Relationalism.[8,9] Given that Spatial Directionalism is consistent both with Spatial Relationalism *and* Spatial Substantivalism, which theory of space *should* the Spatial Directionalist endorse?

To persuade the Spatial Directionalist to side with Spatial Relationalism, some may raise doubts about the coherence of the concept of substantivalist *space*. For example, George Berkeley claimed that he could not conceive of *space*:

116. [P]erhaps, if we inquire narrowly, we shall find we cannot even frame an idea of *pure space*, exclusive of all body. This I must confess seems impossible, as being a most abstract idea. . . . When therefore supposing all the world to be annihilated besides my own body, I say there still remains *pure space*: thereby

nothing else is meant, but only that I conceive it possible, for the limbs of my body to be moved on all sides without the least resistance: but if that too were annihilated, then there could be no motion, and consequently no space.[10]

Space not imaginable by any idea receiv'd from sight, not imaginable, without body moving not even then necessarily existing.[11]

To appease a critic as perplexed as Berkeley, a Spatial Substantivalist may be tempted to elucidate the concept of *space* by noting that regions of *space* are essentially "eternal, uncreated, infinite, indivisible, immutable." As Berkeley himself notes, however, *this* characterization of space is not sufficient to distinguish *space* from God. The Spatial Relationalist, then, may counsel the Spatial Directionalist to resist commitment to substantivalist *space* unless the Spatial Substantivalist offers a coherent account of what *space* would be and of how it would differ from God, properties, states of affairs, and other such entities.

After voicing concern about the coherence of substantivalist *space*, the Spatial Relationalist may appeal to ontological simplicity to press the Spatial Directionalist to side with SR instead of SS. Although both SR and SS imply that there exist both spatial entities and spatial relations, SS, unlike SR, also implies the existence of regions of substantivalist *space*—ontological commitment to mysterious *locations*.

A Spatial Substantivalist may insist that there are good reasons why the Spatial Directionalist should embrace a larger ontology, countenancing the existence of regions of space. In the next chapter, I formulate and evaluate several classic arguments for SS—arguments that turn on plausible intuitions about uniform movement, spatial orientation, uniform expansion, and absolute motion. After explicating the *Leibnizian* Spatial Relationalist's objections to these arguments—objections that turn on rejecting out of hand the plausible spatial intuitions on which these arguments rest—I then explain how, by endorsing Spatial Directionalism, the Spatial Relationalist could instead preserve the Spatial Substantivalist's plausible spatial intuitions *and* reject the Substantivalist's arguments for *space*. Before turning to the formulation and evaluation of classic arguments for SS, consider Spatial Directionalism vis-à-vis Hyperspace—the view that there exist more than three *spatial dimensions*.

IS HYPERSPACE AN OBSTACLE TO GOING NOWHERE?

In his novella, *Flatland*, Edwin A. Abbott makes clear that two-dimensional creatures inhabiting a two-dimensional world would have great difficulty conceiving, and justifying belief in the existence of, a *three*-dimensional material world. Through the voice of A Square, *Flatland*'s two-dimensional narrator, Abbott suggests that three-dimensional creatures—creatures like *us*!—may be similarly handicapped with respect to conceiving a four-dimensional world. Consider an exchange between *Flatland*'s two-dimensional A Square and three-dimensional A Sphere when A Square visits A Sphere's three-dimensional world, Spaceland:

Sphere. But where is this land of Four Dimensions?

. . .

There is no such land. The very idea of it is utterly inconceivable.
[A Square]. Not inconceivable, my Lord . . .
Was I not taught [in Flatland] that when I saw a Line and inferred a Plane, I in reality saw a Third unrecognized Dimension . . . called "height"? And does it not now follow that, in this region, when I see a Plane and infer a Solid, I really see a Fourth unrecognized Dimension . . . ?[12]

Hud Hudson is friendly toward *Hyperspace*—the view that "spacetime is a connected manifold with more than three spatial dimensions . . . [that] can be partitioned into subregions which vary independently with respect to their cosmic conditions." And, a world's cosmic conditions include "facts about the numbers, types, and distributions of existing particles, facts about the relative strengths of the fundamental forces, and facts about the various laws of nature."[13] If Hyperspace is correct, then, just as two-dimensional Flatland would be a subregion of three-dimensional Spaceland, our three-dimensional universe would be a subregion of *the* actual world, which would be a world of *more* than three spatial dimensions—a world that could house hyperspheres, hypercubes, and other more-than-three-dimensional spatial entities.[14] If our three-dimensional universe *could* be home to indefinitely many discrete two-dimensional Flatland-like worlds (stacked, perhaps, one above another), then, for all one knows, hyperspace *could* be home to indefinitely many, and wildly different, three-dimensional universes.

Hudson does not claim to have offered a full-blown defense of Hyperspace, but he does cite the "fine-tuning" of our three-dimensional universe as "very strong evidence" for "the hypothesis of plentitudinous hyperspace." The claim that the universe is finely tuned is the "claim that the range of those values for the cosmic conditions that we have good reason to believe are life-permitting is very small compared with the total range of parameters for which we can reasonably determine whether or not they are life-permitting." Hudson's claim is that the number of combinations of particular laws of nature and values for the physical constants that would allow for the emergence of sentient life is astronomically small compared with the number of combinations of laws of nature and physical-constant values that would *not* allow for the emergence of such life. If Hudson is right, then the existence of a finely tuned three-dimensional universe would be *extraordinarily* unlikely if our universe were the *only* universe that exists or that *has* existed. If, however, there exist *many* three-dimensional universes, then the likelihood would be greater that at least one of them is finely tuned. And, the likelihood that there exist many three-dimensional universes would be greater if Hyperspace is correct—if the actual world *does* involve more than three spatial dimensions, which would allow it to house indefinitely many physically independent *three*-dimensional universes.

Hudson's modest defense of Hyperspace is this: Given that the truth of Hyperspace would make it more likely that multiple, physically independent three-dimensional universes exist, and given that the existence of multiple, physically independent three-dimensional universes would make it more likely that there exists at least one finely tuned universe, the existence of our finely tuned universe constitutes "strong evidence" for Hyperspace. Hudson himself notes that his "fine-tuning" argument does not alone constitute a compelling case for Hyperspace, cautiously concluding that fine-tuning "may render belief in hyperspace not unreasonable" *if* there is "no better evidence for the opposition."[15] To determine whether there *is* a compelling case for Hyperspace, one should weigh the "strong evidence" of "fine-tuning" against (a) objections to Hyperspace itself, (b) the evidence for *competing* explanations for the existence of a finely tuned universe (e.g., the evidence that our finely tuned universe is a function of intelligent design), and (c) the evidence that our universe is *not* a *finely-tuned* universe.

If there does emerge a compelling case for Hyperspace, it would pose no threat to the Directionalist Theory of Space. To understand why, consider first the explanation for why Flatlanders need not abandon a Directional Theory of Space upon discovering that there exists a *third* spatial dimension. The Flatlander who defends DTS would reject substantivalist *space*, claiming that there exist *no* two-dimensional regions of spatial points that are occupied by inhabitants of Flatland. This defender of DTS would then add that all truths that imply the existence of two-dimensional regions of space can be reformulated as truths involving spatial directional relations that clearly do *not* imply that that two-dimensional regions of space exist. The Flatlander could offer the following clarification: "The only spatial directional relations that exist are 'left' and 'right' relations and 'forward' and 'backward' relations. [That is, there exist no 'up' or 'down' relations.] And, every two-dimensional spatial entity involves both 'left' and 'right' spatial directional relations as well as both 'forward' and 'backward' spatial directional relations." If two-dimensional A Square defends DTS and *then* discovers that Flatland is a subregion within a three-dimensional world, then A Square need not abandon

DTS, but simply countenance the existence of indefinitely many additional spatial directional relations—indefinitely many "up" and "down" relations. These would be the relations that Flatland's inhabitants bear to whatever spatial entities exist "above" or "below" Flatland; and they would be the spatial directional relations that obtain among three-dimensional spatial entities.

Three-dimensional defenders of DTS can similarly explain why a compelling case for Hyperspace would pose no threat to Spatial Relationalism. While insisting that all *space*-implying truths are reducible to non-*space*-implying truths involving spatial directional relations, the DTS/Hyperspace defenders can simply acknowledge that there exist more spatial directional relations than they had initially allowed. In addition to countenancing "left," "right," "forward," "backward," "up," and "down" spatial directional relations, DTS/Hyperspace defenders would also countenance—to use the terminology that Hudson adopts—"ana" and "kata" spatial directional relations. The spatial entities in our three-dimensional universe would bear "ana" and "kata" spatial directional relations to any spatial entities that exist outside our three-dimensional universe but within the larger more-than-three-dimensional world of which our three-dimensional universe would be a subworld. The DTS/Hyperspace defender could also add that "ana" and "kata" spatial directional relations are those that obtain among hyperspheres, hypercubes, and other more-than-three-dimensional spatial entities.

WHAT *ARE* SPATIAL DIMENSIONS?

The above explanation of the consistency of DTS with Hyperspace involves much talk of spatial dimensions—talk of Flatland's two spatial dimensions, talk of Spaceland's three dimensions, and talk about Hyperspace's implication that there exist *more* than three spatial dimensions. Strictly, there exists no such thing as a *dimension* that is *had* by a world or that is a *part* of a world. Rather, spatial dimensionality is a way that *all* spatial entities are oriented within a world—a way that a world's spatial entities involve certain kinds of spatial directional relations:[16]

> D3.10 The world is spatially n-dimensional =Df For any time t, there exist n disjoint sets of spatial directional relations, $s_1, \ldots$ and s_n, that are necessarily such that, for any spatial entity x that exists at t, x involves at t at least two spatial directional relations that are members of $s_1, \ldots$ and at least two spatial directional relations that are members of s_n.

Suppose that, at present time T, (a) Flatland is one of multiple two-dimensional universes "embedded" in our three-dimensional universe, (b) there exist multiple two-dimensional universes other than Flatland that are "embedded" in three-dimensional universes other than ours, and (c) all three-dimensional universes (including ours) are "embedded" in *the* four-dimensional world—the world that includes *all* spatial entities. Imagine that, *per impossibile*, you are located outside this four-dimensional world and have a "head-on" outside perspective of *all* spatial entities (but for yourself). There would exist a set S_1 whose members are all and only those relations that are, from your perspective ("parallel") "left/right" spatial directional relations in the sense that if any one of S_1's relations obtain between two spatial entities, one entity would be (from your perspective) to the left of the second while the second would be to the right of the first. There would also exist a set S_2 whose members are all and only those relations that are, from your perspective (parallel) "forward/backward" spatial directional relations; and there would exist sets S_3 and S_4 whose members are, respectively, (e) all and only those spatial directional relations that are, from your perspective, "up/down" and (f) all and only those spatial directional relations that are, from your perspective, "ana/kata."

Assume that, from your imagined outside perspective of *the* four-dimensional world, each two-dimensional universe is stacked above or below another such world and that each two-dimensional universe extends "left/right" and "forward/backward." From your perspective, no "up/down" or "ana/kata" spatial directional relations would obtain between any two of Flatland's spatial entities or between any two of any other two-dimensional universe's spatial entities. "Up/down" or "ana/kata" spatial directional relation *would* obtain between any one of Flatland's spatial entities and a zero-, one-, two-, three- or four-dimensional spatial entity that exists "outside" of Flatland but within *the* four-dimensional world of which, *per impossibile*, you have the outside perspective.

Relative to your outside perspective at T, "left/right," "forward/backward," *and* "up/down" spatial directional relations *would* obtain between any two spatial entities within our three-dimensional universe; but among our universe's spatial entities, *no* "ana/kata" spatial directional relation would obtain. From your outside perspective, however, you *could* be aware that a spatial entity within our three-dimensional universe bears "ana/kata" spatial directional relations to spatial entities that exist outside our universe but within the four-dimensional world of which you have the outside perspective.

Although Spatial Directionalism implies that there are objective facts about which spatial directional relations do and do not obtain between any two spatial entities, the directionalist account of dimensionality that D12 captures does *not* imply that there is a "correct" or "privileged" conception or perspective of how spatial entities are oriented with respect to "the dimensions" of the universe. For example, in a world of four spatial dimensions, any given spatial entity involves at least two of each of four sets of spatial directional relations, but it need not also be true that one of these four sets corresponds with "absolute *up*" or "absolute *north*." Accepting the directionalist concept of dimensionality is consistent with rejecting the view that every spatial entity has an objective orientation vis-à-vis "absolute up and down," "absolute left and right," "absolute forward and backward," and "absolute ana and kata."

To make this clearer, assume that the actual world *is* a world of four spatial dimensions and that, on three-dimensional Earth, Jenny the Scout has a metaphysical compass equipped with a digital display: At any given time, the compass's red-tipped needle points toward Earth's North Magnetic Pole and the digital display reveals the exact spatial directional relation that the needle's red tip then bears to its opposite non-red tip. Imagine that, when you observe Jenny's compass from your head-on perspective outside *the* four-dimensional world at time T, you observe that the compass needle's red tip points toward three-dimensional Earth's North Magnetic Pole and that the red tip bears the spatial directional relation d_{489} to the needle's opposite non-red tip. If, from your head-on outside perspective, the spatial directional relations that you would conceive to be "left/right," "forward/backward," "up/down," and "ana/kata" relations are members of, respectively, sets S_1, S_2, S_3, and S_4, then relation d_{489} would be a member of S_3.

Imagine now that, while there is no change among the spatial directional relations that obtain among any of the four-dimensional world's spatial entities, you float upward until you have an outside perspective that is at a 45° angle to the four-dimensional world. You would continue to observe that Jenny's metaphysical compass needle points toward Earth's North Magnetic Pole and that the compass's digital display continues to indicate that the needle's red tip bears the spatial directional relation d_{489} to its non-red tip. Also, the compass needle would continue to involve at least two members of each S_1, S_2, S_3, and S_4; and d_{489} would continue to be a member of S_3. From your new angled outside perspective, however, you will no longer consider the spatial directional relations that are members of sets S_1, S_2, S_3, and S_4 to be the relations that correspond with those relations that you now consider to be "left/right," "forward/backward," "up/down," and "ana/kata." Rather, from your new angled outside perspective, the spatial directional relations that you now consider to be "left/right," "forward/backward," "up/down," and "ana/kata" will be those that are the members of four other sets—those that are the members of, respectively, say, sets S_5, S_6, S_7, and S_8. The compass needle *is* oriented in a four-dimensional world in the sense that it involves at least two members of each S_1, S_2, S_3, and S_4 and also of S_5, S_6, S_7, and S_8; but neither Spatial Directionalism nor the directionalist account of dimensionality implies that S_1-S_4 or S_5-S_8 or some other class of sets corresponds with "absolute left/right," "absolute forward/backward," "absolute up/down," and "absolute ana/kata." Put another way, neither Spatial Directionalism nor the directionalist account of dimensionality implies that there is a "privileged" or "correct" outside perspective with respect to whether d_{489} is a "left/right," "forward/backward," "up/down," or "ana/kata" spatial directional relation.

Summary. There are, then, reasons to endorse the Directionalist Theory of Space. First, by allowing the reduction of *space*-implying claims to directionalist/relationalist claims that do not imply the existence of substantivalist *space*, DTS allows one to sidestep any metaphysical mysteries that commitment to regions of space or spatial points may involve. Second, DTS offers an ontological economy that Spatial Substantivalism does not. Third, DTS is consistent with Hyperspace should there emerge compelling philosophical or scientific reasons for endorsing the existence of more than three spatial dimensions.[17] Finally, the Spatial Directionalism that underlies DTS allows one to explain spatial dimensionality in terms of spatial directional relations. Consider now four classic arguments on behalf of Spatial Substantivalism—four reasons to posit the existence of substantivalist *space*.

NOTES

1. L.V.47, *The Leibniz-Clarke Correspondence*, pp. 69–70.
2. Cartwright, "Scattered Objects," p. 153.
3. See Cartwright, "Scattered Objects," p. 153. Cartwright claims that no region "that consists of a single point [of space] or, for that matter, of any finite number of points" can be a receptacle for a material object. Apparently, Cartwright would deny the existence of *part*-less material entities and of scattered objects composed solely of *part*-less material entities.
4. D3.1 and D3.2 not only allow for the possibility that mereological simples exist, but they also allow for the possibility that a single mereological simple is the *only* spatial entity that exists. In a stark universe in which mereological simple MS is the only spatial entity at a time T, MS would bear *no* spatial directional relation to anything at T, but MS would nonetheless have an actual spatial location: Where P is a possible location for MS, MS and P at T are such that, at T, *if* there exists something to which MS bears spatial directional relation D at T, then D would be a member of P.
5. See Markosian, "A Spatial Approach to Mereology," p. 73: "[T]hat a part must occupy a subregion of the region occupied by the whole is a non-negotiable feature of our everyday conception of a physical object," and "the idea that occupying a subregion is a *sufficient* condition for being a part has a great deal of intuitive appeal." Cf. Def.8, Schol.III, *Newton's Mathematical Principles*, p. 7: "[T]he place of the whole is the same as the sum of the places of the parts."
6. As Roderick M. Chisholm has noted, the definition is not needlessly complicated: The third clause allows for the possibility that monads exist and that a monad can be a proper part of one of two objects that compose a whole. See *Person and Object: A Metaphysical Study*, pp. 152–53.
7. L.V.47, *The Leibniz-Clarke Correspondence*, pp. 69–70.

8. See Michael Friedman, *Foundations of Space-Time Theories: Relativistic Physics and Philosophy of Science* (Princeton, NJ: Princeton University Press, 1983), pp. 62–63; see also pp. 217, 221. Friedman writes that "[t]he Leibnizian relationalist is perfectly happy to take spatio-temporal properties and relations as primitive," noting that the "primitive relations of distance, simultaneity, or temporal precedence [can hold] between physical events, but such relations never hold between unoccupied space-time points." If DTS is correct, then the relationalist should acknowledge that spatial directional relations are among those primitive relations that can obtain between physical entities, but never between points of space or spacetime.

9. See Hartry Field, "Can We Dispense With Space-Time?" *PSA: Proceedings of the Biennial Meeting of the Philosophy of Science Association* (1984), 47–48. Field claims that relationalists violate "the whole spirit of relationalism" if they appeal to "heavy duty platonism"—the view that relations obtain between physical things and (abstract) numbers. Such relationalists, Field complains, abandon the view that one can do without spacetime by relying instead on "*relations between aggregates of matter.*" DTS, as formulated, *is* platonist, but it remains true to the spirit of relationalism by relying on (directionalist) relations that obtain among spatial entities, not between spatial entities and abstracta.

10. Berkeley, 116, *A Treatise Concerning the Principles of Human Knowledge*, p. 93.

11. Berkeley, 135, *Philosophical Commentaries*, Notebook B, ed. A. A. Luce, *The Works of George Berkeley, Bishop of Cloyne*, ed. Luce and T. E. Jessop, Vol. I (London: Thomas Nelson and Sons Ltd., 1951), p. 20.

12. Edwin A. Abbott, *Flatland: A Romance of Many Dimensions*, 2nd ed. (New York: Dover Publications, 1952), pp. 87–88.

13. Hud Hudson, *The Metaphysics of Hyperspace* (Oxford and New York: Oxford University Press, 2005), p. 40.

14. For assistance with conceptualizing worlds with fewer or more than three spatial dimensions, see Thomas Banchoff, *Beyond the Third Dimension*, Scientific American Library Series (New York: W. H. Freeman & Co., 1990).

15. Hudson, *The Metaphysics of Hyperspace*, p. 41.

16. Cf. Brian Greene, *The Fabric of the Cosmos: Space, Time, and the Texture of Reality* (New York: Vintage Books, 2005), p. 360: "[W]hen we say that there are three space dimensions we mean that there are three independent directions or axes along which you can move." Greene also writes that if the universe *does* have an additional spatial dimension heretofore undetected, then there would be "another independent direction in which things can move."

17. Physicists who defend string theory posit the existence of at least nine spatial dimensions. For an introduction to string theory and its implications for spatial dimensionality, see Greene, *The Fabric of the Cosmos*, pp. 327–75.

Chapter 4

Defending Spacelessness

This chapter includes the formulation and evaluation of four classic arguments on behalf of Spatial Substantivalism—arguments that turn on plausible claims involving the motion, spatial orientation, and size of spatial entities. Leibniz responded directly to two of these arguments in his correspondence with Samuel Clarke;[1] Poincaré entertained (and rejected) the third, and Newton defended an argument close to the fourth. After formulating each argument, I explicate the objection that a Leibnizian Spatial Relationalist would offer—an objection that involves categorically rejecting the plausible spatial intuition on which the substantivalist argument turns. With respect to each of these four arguments, I then explain how, by embracing spatial directional relations, the Spatial Directionalist can also object to the argument but *without* rejecting the plausible spatial intuition that underlies the argument. Following the evaluation of these arguments on behalf of SS, I address an objection to the presupposition that spatial directional relations exist—a presupposition that is essential to the success of the directionalist objections to the four substantivalist arguments.

Whether the Spatial Directionalist who objects to the four substantivalist arguments should thereby object to Spatial Substantivalism and endorse Spatial Relationalism remains an open question until this chapter's final section. In the final section, I sketch a modest defense of the Directional Theory of Space—the result of coupling Spatial Relationalism with Spatial Directionalism.

THE UNIFORM MOTION ARGUMENT FOR SS

Clarke argues that one should embrace substantivalist (absolute) *space* on grounds that Spatial Relationalism cannot allow for the possibility that *all* spatial entities move uniformly in the same direction:

> If space was nothing but the order of things coexisting; it would follow, that if God should remove in a straight line the whole material world entire, with any swiftness whatsoever; yet it would still always continue in the same place.[2]

Informally, the Uniform Motion Argument is this: "It is at least possible that every existing spatial entity moves simultaneously in the same direction at the same velocity. For example, if there exists a triangular array of marbles, and if *everything* moves uniformly, then the triangular array would remain intact: The marbles would remain at exactly the same distance from one another and from any other spatial entities; and the angles that obtain among any three marbles (or other spatial entities) would remain unchanged. In short, then, if all spatial entities move uniformly, then all spatial entities would continue to bear the same spatial relations to one another throughout the period of uniform movement. This possibility poses a problem for Spatial Relationalism. If the Spatial Relationalist is right that substantivalist space does not exist, then a spatial entity's movement *cannot* involve its moving from one (substantivalist) *place* to another; rather, a spatial entity's movement would involve a change in the spatial relations that the entity bears to other spatial entities. But if spatial entities move only if there is a change of the spatial relations that they bear to other entities, then it is impossible that all spatial entities move *uniformly*: Such uniform motion would require that the array of all spatial entities moves *intact*—that the spatial relations that obtain among the various entities do *not* change as they move. Thus, that uniform motion *is* possible implies that SR is not correct given that SR implies that uniform motion is *im*possible. To reject SR is to endorse Spatial Substantivalism, and SS *does* allow for the possibility that all spatial entities move uniformly: If substantivalist space exists, then uniform motion would involve *all* spatial entities occupying successively different regions of *space* throughout the interval of motion while, throughout that interval, the spatial relations that obtain among those entities remain unchanged. Put another way, substantivalist uniform motion would involve each spatial entity's changing the spatial relations that it bears to various *regions of space* without changing the spatial relations that it bears to other spatial entities."[3]

More formally, the Uniform Motion Argument for SS is this:

UM 1. It is possible that every spatial entity moves uniformly between times t_1 and t_2.
   2. If it is possible that every spatial entity moves uniformly between times t_1 and t_2, then it is possible that every spatial entity occupies a different place at each time between t_1 and t_2 and that the spatial relations that obtain between any two spatial entities at any time between t_1 and t_2 are the same spatial relations that obtain between those two entities at any other time between t_1 and t_2.
   3. Therefore, it is possible that every spatial entity occupies a different place at each time between t_1 and t_2 and that the spatial relations that obtain between any two spatial entities at any time between t_1 and t_2 are the same spatial relations that obtain between those two entities at any other time between t_1 and t_2. (from UM-1, UM-2)
   4. Therefore, if SR is correct, then it is possible that every spatial entity occupies a different place at each time between t_1 and t_2 and that the spatial relations that obtain between any two spatial entities at any time between t_1 and t_2 are the same spatial relations that obtain between those two entities at any other time between t_1 and t_2. (from UM-3)
   5. Therefore, if SR is correct, then it is possible that every spatial entity occupies a different place at each time between t_1 and t_2. (from UM-4)
   6. If SR is correct only if it is possible that every spatial entity occupies a different place at each time between t_1 and t_2, then SR is correct only if it possible that every spatial entity is such that the spatial relations that it bears to other spatial entities at any given time between t_1 and t_2 are *other* than those that it bears to other spatial entities at any other time between t_1 and t_2.
   7. Therefore, SR is correct only if it possible that every spatial entity is such that the spatial relations that it bears to other spatial entities at any given time between t_1 and t_2 are *other* than those that it bears to other spatial entities at any other time between t_1 and t_2. (from UM-5, UM-6)
   8. Therefore, if SR is correct, then it is possible that the spatial relations that obtain between any two spatial entities at any time between t_1 and t_2 are the *same* spatial relations that obtain between those two entities at any other time between t_1 and t_2. (from UM-4)
   9. Therefore, SR is correct only if it is possible that every spatial entity is such that the spatial relations that it bears to other spatial entities at any given time between t_1 and t_2 are *other* than those that it bears to other spatial entities at any other time between t_1 and t_2 and that the spatial relations that obtain between any two spatial entities at any time between t_1 and t_2

are the *same* spatial relations that obtain between those two entities at any other time between t_1 and t_2. (from UM-7, UM-8)
10. Therefore, it is not possible that SR is correct. (from UM-9)
11. If SR is not correct, then Spatial Substantivalism *is* correct.
12. Therefore, SS is correct. (from UM-10, UM-11)

The Leibnizian would counsel Spatial Relationalists to endorse the second premise, noting that it is nothing more than a statement of what uniform motion would involve: All spatial entities move uniformly throughout an interval, if and only if, they successively occupy different places while retaining their same spatial arrangement (i.e., while each entity continues to bear the same spatial relations to other entities).

The Leibnizian would also counsel Spatial Relationalists to endorse the sixth premise: *If* SR is correct only if it allows for the possibility that all spatial entities successively occupy different places throughout an interval, then it is possible that the spatial relations that obtain among all spatial entities at a given time are other than those that obtain at other times throughout the interval of uniform movement. After all, the Leibnizian would remind us, motion does *not* involve a spatial entity's moving from one region of substantivalist space to another; rather, motion involves a spatial entity's moving to a different *place*, which *is* the changing of spatial relations that that entity bears to other spatial entities:

> When it happens that one of those co-existent things changes its relation to a multitude of others, which do not change their relation among themselves; and that another thing, newly come, acquires the same relation to the others, as the former had; we then say, it is come into the place of the former; and this change, we call a motion in that body. . . . And, to give a kind of a definition: *place* is that, which we say is the same to A and, to B, when the relation of the co-existence of B, with C, E, F, G, etc. agrees perfectly with the relations of the co-existence, which A had with the same C, E, F, G, etc.[4]

Aware that one could reduce SR to absurdity by coupling UM-2 and UM-6 with the possibility that all spatial entities move uniformly, the Leibnizian would *reject* UM-1, claiming that it is *im*possible that all spatial entities move uniformly: "I have demonstrated, that space is nothing else but an order of the existence of things, observed as existing together; and therefore the fiction of a material finite universe, moving forward in an infinite empty space, cannot be admitted."[5] By rejecting UM-1, the Leibnizian Spatial Relationalist would thereby block Clarke's inference to Spatial Substantivalism.

There is an alternative objection available to the Spatial Directionalist—an alternative that would allow the Spatial Directionalist to reject the UM defense of SS while preserving the possibility that all spatial entities undergo uniform motion. In short, by appealing to spatial *directional* relations, the Spatial Directionalist could *defend* UM-6 *and* the possibility of uniform motion (UM-1), blocking the inference to SS by rejecting UM-2. Here's how.

First, such a Spatial Directionalist would agree with Leibnizian Spatial Relationalists that UM-6 is true: *If* every spatial entity undergoes uniform motion, then every entity *would* change its place; and, per SR, an entity's changing its place *is* that entity's changing the spatial relations that it bears to other entities. Thus, the Spatial Directionalist would agree with Leibnizian Spatial Relationalists that if *all* spatial entities move uniformly, then every entity *would* change the spatial relations that it bears to others.

Second, allowing for the possibility that all spatial entities undergo uniform motion, the Spatial Directionalist would *reject* UM-2, which is the Leibnizian's presupposition that, if every spatial entity undergoes *uniform* motion, then the spatial relations that obtain among spatial entities would *not* change. The Spatial Directionalist would then argue that *if* every spatial entity undergoes uniform motion in a *directioned* universe—in a universe that includes spatial *directional* relations—then the spatial *directional* relations that would obtain between any two spatial entities *would* change successively during an interval of uniform motion.

For example, suppose that all spatial entities move uniformly between Monday and Friday and that these entities include (as represented in figure 4.1) a triangular array of six marbles 5 cm apart. Although Leibniz would note that marbles #2 and #3 both bear, say, the spatial relation *is-5-cm-at-a-45°-angle-from* to, respectively, marbles #4 and #5 throughout the interval of uniform motion, DTS implies that that marbles #2 and #3 would each successively bear different spatial *directional* relations to marbles #4 and #5 throughout this interval. For example, at time *t*, marble #2 may bear the

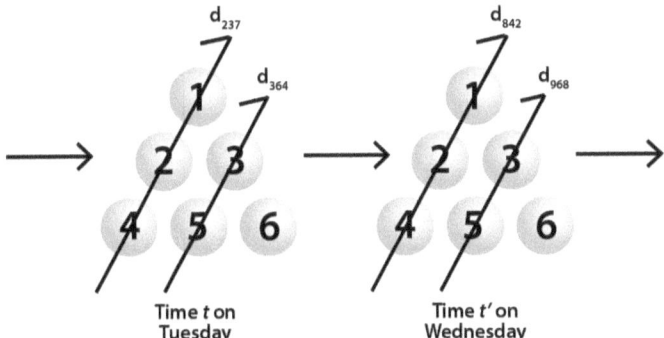

Figure 4.1 Uniform Motion per Spatial Directionalism. *Source*: C. Frantom, J. Rich: White Roche LLC.

spatial directional relation *is* d_{237} *of* to #4 while #3 bears *is* d_{364} *of* to #5; then, at *t'*, #2 bears *is* d_{842} *of* to #4 and #3 bears *is* d_{968} *of* to #5. The implication, then, is that the spatial directional relations that any uniformly moving spatial entity *involves* would change from moment to moment throughout its interval of uniform motion. Thus, by affirming that uniform motion *does* involve the successive changing of the spatial *directional* relations that obtain among the spatial entities moving uniformly, the Spatial Directionalist can both reject the Uniform Motion Argument for SS *and* preserve the plausible Clarke/Newton intuition that, possibly, all spatial entities do move uniformly.[6]

Brentano charged that "Leibniz . . . was completely mistaken when he made spatial location something entirely relative, so that something could go from one place to another, without changing in any way at all."[7] If the Spatial Directionalist can be persuaded to endorse the Directional Theory of Space by coupling Spatial Directionalism with Spatial Relationalism, then this defender of DTS would agree with Brentano that nothing can change places without undergoing *some* kind of spatial change. But this defender of DTS would adamantly reject Brentano's assessment of Leibniz, noting that Leibniz was *not* mistaken "when he made spatial location something entirely relative." The defender of DTS—the Spatial Relationalist who also accepts Spatial Directionalism—would insist that a spatial entity's location *is* "entirely relative": A spatial entity's location *is* the set of spatial *directional* relations that it involves.

THE INVERTED WORLD ARGUMENT FOR SS

In the correspondence with Leibniz, Clarke addresses the Principle of Sufficient Reason vis-à-vis the will of God:

> nothing is, without a sufficient reason why it is, and why it is thus rather than otherwise. . . . But this sufficient reason is oft-times no other, than the mere will of God. For instance: why this particular system of matter, should be created in one particular place, and that in another particular place; when, (all place being absolutely indifferent to all matter,) it would have been exactly the same thing *vice versa*, supposing the two systems . . . of matter to be alike; there can be no other reason, but the mere will of God.[8]

One may be tempted to construe Clarke's remarks as a version of the Inverted World Argument for SS:

> The actual world includes spatial entities in a particular array—spatial entities that bear certain spatial relations to one another in a particular spatial ordering. There is, however, a nonactual possible world that is the exact inversion of the actual

Defending Spacelessness 55

world: All spatial entities bear exactly the same spatial relations to one another, but in a spatial ordering that is the mirror-image of their ordering in the actual world. Consider again the triangular array of marbles represented in figure 4.1. Marbles #2 and #3 would bear the spatial relation *is-5-cm-at-a-45°-angle-from* to, respectively, marbles #4 and #5 in both the actual world *and* in the inverted world. And, in both worlds, #5 would bear the relation *is-5-cm-to-the-side-of* to #4 and to #6. That the inverted world is distinct from the actual world implies that substantivalist space exists: For any two spatial entities, there would be no difference in the spatial relations that obtain between them in the actual world and in the inverted world; so, the distinction between the two worlds can only be explained in terms of the different regions of substantivalist space that the spatial entities would occupy in the two worlds. For example, although marble #5 may be 5 cm away from #4 and away from #6 in both the actual and the inverted worlds, the region of space occupied by #4 in the actual world would be the region occupied by #6 in the inverted world and *vice versa*. Thus, only by endorsing SS can one allow for the distinction between the actual and the inverted worlds: Given that the actual and inverted worlds *are* distinct possible worlds that are equivalent with respect to the spatial relations that obtain, the worlds would differ only with respect to the regions of substantivalist space occupied by the spatial entities that inhabit both worlds.[9]

Below is a formulation of the Inverted World Defense of SS:

IW 1. There is a possible world w^* that is the inverted, mirror-image counterpart of actual world w.
 2. If there is a possible world w^* that is the inverted, mirror-image counterpart of actual world w, then (a) for any time t, the spatial relations that obtain between any two spatial entities at t in w^* are the same spatial relations that obtain between those two entities at t in w, and (b) w^* is not identical with w.
 3. Therefore, (a) for any time t, the spatial relations that obtain between any two spatial entities at t in w^* are the same spatial relations that obtain between those two entities at t in w, and (b) w^* is not identical with w. (from IW-1, IW-2)
 4. Therefore, if SR is correct, then (a) for any time t, the spatial relations that obtain between any two spatial entities at t in w^* are the same spatial relations that obtain between those two entities at t in w, and (b) w^* is not identical with w. (from IW-3)
 5. Therefore, SR is correct only if, for any time t, the spatial relations that obtain between any two spatial entities at t in w^* are the same spatial relations that obtain between those two entities at t in w. (from IW-4)
 6. If SR is correct only if, for any time t, the spatial relations that obtain between any two spatial entities at t in w^* are the same spatial relations

that obtain between those two entities at *t* in *w*, then SR is correct only if *w** is identical with *w*.
7. Therefore, SR is correct only if *w** is identical with *w*. (from IW-5, IW-6)
8. Therefore, SR is correct only if *w** is not identical with *w*. (from IW-4)
9. Therefore, SR is correct only if *w** is both identical and not identical with *w*. (from IW-7, IW-8)
10. Therefore, SR is not correct. (from IW-9)
11. If SR is not correct, then Spatial Substantivalism is correct.
12. Therefore, SS is correct. (from IW-10, IW-11)

Without affirming its antecedent, the Leibnizian Spatial Relationalist should agree with conditional IW-2: *If* there is a possible world that *is* the inverted counterpart of the actual world, then there *would* be two distinct possible worlds even though the same spatial relations would obtain among the same spatial entities in both worlds. The Leibnizian Spatial Relationalist should also endorse IW-6: If certain spatial relations obtain among the existing spatial entities in a given possible world, then that world *would* be identical with any possible world in which all and only those same spatial relations obtain among the same spatial entities.

So, to block the inference to Spatial Substantivalism, the Leibnizian Spatial Relationalist should object to IW-1. After formulating and evaluating Leibniz's own objection to IW-1 (along with Clarke's rebuttal), I will explain how the Spatial Directionalist could instead object to the Inverted World Argument while *preserving* IW-1—while preserving the plausible intuition that the inverted counterpart of the actual world could have existed instead.

Leibniz objects to IW-1 by invoking his Principle of Sufficient Reason [PSR]—the principle that, for any true proposition, there is an explanation for why it is true rather than false:[10]

> [I]f space was an absolute being, there would something happen for which it would be impossible there should be a sufficient reason. Which is against my axiom. And prove it thus. Space is something absolutely uniform; and, without the things placed in it, one point of space does not absolutely differ in any respect whatsoever from another point of space. Now from hence it follows, (supposing space to be something in itself, besides the order of bodies among themselves,) that 'tis impossible there should be a reason, why God, preserving the same situation of bodies among themselves, should have placed them in space after one certain particular manner, and not otherwise; why every thing was not placed the quite contrary way, for instance, by changing East into West. But if space is nothing else, but that order or relation; and is nothing at all without bodies, but the possibility of placing them; then those two states, the one

Defending Spacelessness 57

such as it now is, the other supposed to be the quite contrary way, would not at all differ from one another. Their difference therefore is only to be found in our chimerical supposition of the reality of space in itself. But in truth the one would exactly be the same thing as the other, they being absolutely indiscernible; and consequently there is no room to enquire after a reason of the preference of the one to the other.[11]

Consider a paraphrase of Leibniz's objection to IW-1:

If w and w^* *are* two distinct possible worlds such that w^* is the inverted version of w, then God *could* have created w or w^*. If God could have created either possible world, then, given the Principle of Sufficient Reason, there would have been *some* explanation for why God chose to create w instead of w^*. But there could have been no sufficient reason for God's creating one world rather than the other. After all, the spatial relations that obtain among things in the actual world are the *same* spatial relations that would obtain among those *same* things in w^*: Any two spatial entities that would exist at a certain distance and angle from one another in w would also exist at that same distance and angle from one another in w^*. Because w and w^* would thereby be indiscernible, there could have been no difference between w and w^* that could have explained why God chose to create w instead of w^*! And, positing the existence of points of substantivalist *space* would be of no help in preserving IW-1: Because there would be no discernible difference between any two points of space, there could be no discernible difference between a spatial entity's occupying certain points of space in w and certain other points of space in w^*. And, if there could be no discernible difference between a spatial entity's occupying certain points of space in w and certain other points of space in w^*, then there could have been no difference between w and w^* that could have explained why God chose to create w rather than w^*. The implication, then, is this: (a) IW-1 is true only if w and w^* are distinct possible worlds, (b) w and w^* are distinct worlds only if God could have had a sufficient reason for creating w rather than w^*, (c) God could have had *no* such reason given that w and w^* are indiscernible, so (d) IW-1 is false—w and w^* are *not* distinct possible worlds.[12]

Below is a more formal interpretation of Leibniz's objection to IW-1:

NIW 1. For any proposition that is true, there is some sufficient explanation for why it is true rather than false. [Principle of Sufficient Reason]
   2. If PSR is correct and if there is a possible world that is the inverted, mirror-image counterpart of the actual world, then there could be an explanation of why the actual spatial ordering (instead of the inverted ordering) is the actual spatial ordering of things.

3. Therefore, if there is a possible world that is the inverted, mirror-image counterpart of the actual world, then there could be an explanation of why the actual spatial ordering (instead of the inverted ordering) is the actual spatial ordering of things. (from NIW-1, NIW-2)
4. There couldn't possibly be an explanation for why the actual ordering (rather than the inverted ordering) is actual.
5. Therefore, it is false that there is a possible world that is the inverted, mirror-image counterpart of the actual world. (from NIW-3, NIW-4)

To rebut Leibniz's objection, Clarke rejects the Principle of Sufficient Reason [NIW-1], claiming that God could well have created either the actual world *or* its "perfectly equal" inverted counterpart even though, with respect to those two indiscernible worlds, there would have been *no* explanation for why God created one rather than the other:

> there could be no other reason but mere will, why three equal particles should be placed or ranged in the order *a*, *b*, *c*, rather than in the contrary order. And therefore no argument can be drawn from this indifferency of all places, to prove that no space is real.[13]

> God's placing one cube of matter behind another equal cube of matter, rather than the other behind that; is a choice no way unworthy of the perfections of God, though both these situations be perfectly equal.[14]

> If it is possible for God to make or to have made two pieces of matter exactly alike, so that the transposing in situation would be perfectly indifferent; this learned author's notion of a sufficient reason falls to the ground.[15]

According to Clarke, God could have made any number of different choices involving indiscernible worlds and thereby could have made choices without there having been a sufficient reason for making one choice rather than the other. Thus, Clarke concludes, the Principle of Sufficient Reason is false: There *are* true statements about God's decisions, and there are no reasons that explain why those statements are true rather than false. Regardless of whether Clarke's rebuttal constitutes a definitive objection to NIW-1, Clarke's remarks make clear that PSR is at best controversial. Given that the Inverted World Argument for SS turns, in part, on the plausible intuition that the actual world's inverted counterpart is a distinct possible world that could have existed instead, the Spatial Relationalist needs an objection to IW that is more compelling than Leibniz's appeal to PSR. Spatial Directionalism affords the Spatial Relationalist just such an objection.

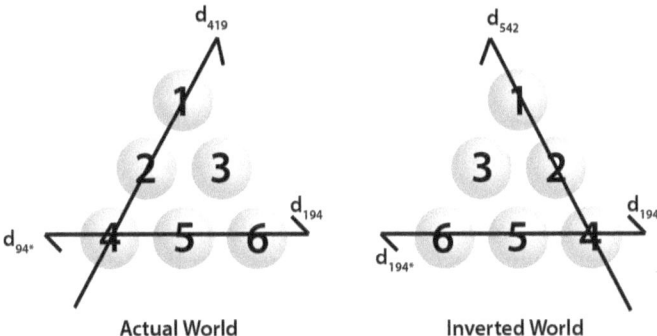

Figure 4.2 An Inverted World per Spatial Directionalism. *Source*: C. Frantom, J. Rich: White Roche LLC.

Taking no stand on whether Spatial Relationalism is correct, the Spatial Directionalist can both object to IW *and* preserve Clarke's plausible intuition that the actual world's inverted counterpart could have existed instead. Without denying IW-1, the Spatial Directionalist could object to IW-2, claiming that, with respect to the inverted possible world, it is *false* that the spatial relations that would obtain between any two spatial entities would be the same that obtain between them in the actual world. In the actual *directioned* universe at a given time, any spatial entity that actually bears certain spatial directional relations to some other spatial entity would, in the inverted world, bear *none* of those relations to that other entity; it would instead bear the *opposite* spatial directional relations to that entity. Consider again a triangular array of six marbles (represented in figure 4.2) and compare their orientation in the actual world and in the actual world's inverted counterpart. Assume that marble #6 bears spatial directional relation d_{194} to marbles #5 and #4 in the actual world, but not in the inverted world; in the inverted world, marble #6 would bear opposite relation d_{194*} (but not d_{194}) to marbles #5 and #4. Marble #1 may bear spatial directional relation d_{419} to #4 in the actual world, and it may bear an altogether different relation to #4 in the inverted world. This example makes clear that, by invoking spatial directional relations, the Spatial Directionalist can reject IW-2, claiming that the spatial relations that a spatial entity involves in one possible world be *other* than the relations it involves in that possible world's inverted counterpart.

The Spatial Directionalist who defends IW-1 and rejects IW-2 objects to Clarke's view that the existence of substantivalist *space* is a necessary condition for the possibility that a world has an inverted counterpart. This objection to Clarke is not itself an argument *for* Spatial Relationalism, but instead leaves open the question of whether the Spatial Directionalist should side with Spatial Relationalism or with Spatial Substantivalism, which is a matter addressed in this chapter's final section.

THE UNIFORM EXPANSION ARGUMENT FOR SS

Brentano insists that adequate views regarding space and time must allow that there *is* "absolute magnitude" with respect to that which has temporal duration or spatial extension:

> If I know that someone has twice as large a fortune as another then I may be ignorant of the absolute size of his fortune, but that it has some determinate size is clear to me none the less. . . . For the fortune that is twice as large can be owned just as well by someone who possess millions, thousands or only hundreds. The most various of absolute specifications are conceivable, but the absence of every specific absolute magnitude quite inconceivable. Thus the assertions on the part of the physicists of the mere relativity of what is spatial and temporal have always appeared to me to be grotesque.[16]

Henri Poincaré suggests, but does *not* defend, a third defense of Spatial Substantivalism that trades on the view that Brentano endorses:

> Suppose that in one night all the dimensions of the universe became a thousand times larger. The world will remain *similar* to itself, if we give the word *similitude* the meaning it has in the third book of Euclid. Only, what was formerly a metre long will now measure a kilometer, and what was a millimeter long will become a metre. The bed in which I went to sleep and my body itself will have grown in the same proportion. When I wake in the morning what will be my feeling in face of such an astonishing transformation? Well, I shall not notice anything at all. The most exact measures will be incapable of revealing anything of this tremendous change, since the yard-measures I shall use will have varied in exactly the same proportions as the objects I shall attempt to measure. In reality the change only exists for those who argue as if space were absolute.[17]

Informally, the Uniform Expansion Argument (which Poincaré explicates *and* rejects) is this:

> It is at least possible that, at a given time, every existing non-zero-dimensional spatial entity doubles or triples in size regardless of whether this could be empirically verified. (After all, if *everything* uniformly doubles, then metersticks, tape measures, compasses, and protractors would all double as well!)

After doubling in size, however, existing things would continue to bear the *same* spatial relations to one another: A chair that is one meterstick away from a table at such-and-such an angle would remain a meterstick away at the same angle after the chair, table, *and* meterstick, double in size. If such uniform expansion *is* a possibility, then, spatially, there is no relational difference between the unexpanded world and the expanded world. Thus, the spatial difference between the unexpanded world and expanded world would involve substantivalist *space*: The region of *space* occupied by any given object in the expanded world would be twice the size of the region of *space* that it occupies in the unexpanded world.

More formally, the Uniform Expansion Argument for SS is this:

UE 1. There is a possible world w in which every spatial entity doubles in size between t_1 and t_2.
 2. If there is a possible world w in which every spatial entity uniformly doubles in size between t_1 and t_2, then (i) it is possible that, at t_2, every spatial entity in w occupies a larger place than it occupied at t_1 and (ii) it is possible that, at t_2, every spatial entity in w bears to other spatial entities the same spatial relations that it bore to those entities at t_1.
 3. Therefore, (i) it is possible that, at t_2, every spatial entity in w occupies a larger place than it occupied at t_1 and (ii) it is possible that, at t_2, every spatial entity in w bears to other spatial entities the same spatial relations that it bore to those entities at t_1. (from UE-1, UE-2)
 4. Therefore, if Spatial Relationalism is correct, then (i) it is possible that, at t_2, every spatial entity in w occupies a larger place than it occupied at t_1 and (ii) it is possible that, at t_2, every spatial entity in w bears to other spatial entities the same spatial relations that it bore to those entities at t_1. (from UE-3)
 5. Therefore, if SR is correct, then it is possible that, at t_2, every spatial entity in w bears to other spatial entities the same spatial relations that it bore to those entities at t_1. (from UE-4)
 6. Therefore, if Spatial Relationalism is correct, then it is possible that, at t_2, every spatial entity in w occupies a larger place than it occupied at t_1. (from UE-4)
 7. If it is possible that, at t_2, every spatial entity in w occupies a larger place than it occupied at t_1, then it is possible that, at t_2, every spatial entity in w fails to bear to other spatial entities the same spatial relations that it bore to those entities at t_1.
 8. Therefore, if Spatial Relationalism is correct, then it is possible that, at t_2, every spatial entity in w fails to bear to other spatial entities the same spatial relations that it bore to those entities at t_1. (from UE-6, UE-7)

9. Therefore, if SR is correct, then it is possible that, at t_2, every spatial entity in w both bears and fails to bear to other spatial entities the same spatial relations that it bore to those entities at t_1. (from UE-5, UE-8)
10. Therefore, it is not possible that SR is correct. (from UE-9)
11. If SR is not correct, then Spatial Substantivalism is correct.
12. Therefore, SS is correct. (from UE-10, UE-11)

Poincaré does not defend UE. Rather, after noting that SS *does* imply the possibility of uniform expansion, Poincaré affirms SR and thereby concludes, as would Leibnizian Spatial Relationalists, that uniform expansion is *impossible*: "[I]t would be better to say that as space is relative, nothing at all [i.e., no uniform expansion] has happened, and that it is for that reason that we have noticed nothing [i.e., have observed *no* uniform expansion]."[18] In terms of UE, then, Poincaré would flatly reject UE-1.

Alternatively, however, a Spatial Directionalist can resist UE by appealing to spatial directional relations to reject UE-2 while both allowing for the possibility of uniform expansion *and* leaving open the question of whether SR or SS is correct. To reject UE-2, the Spatial Directionalist will claim that the possibility of uniform expansion does *not* imply that spatial entities, after uniform expansion, would bear to other entities the same spatial relations that they would have born to those entities before the expansion.

For example, imagine a materially solid cube C (represented in figure 4.3) that is composed of W, X, Y, and Z. Imagine as well that, just before the universe undergoes uniform expansion, X bears spatial directional relation d_{364} to Z and that relation d_{593} does not obtain at all. Now imagine that the universe

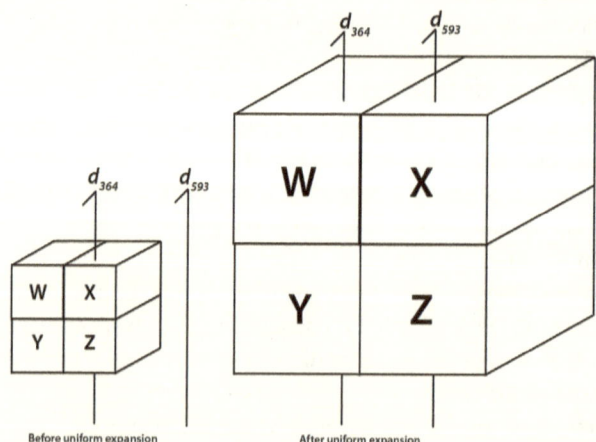

Figure 4.3 Uniform Expansion per Spatial Directionalism. *Source*: C. Frantom, J. Rich: White Roche LLC.

undergoes uniform expansion such that all non-zero-dimensional spatial entities expand uniformly in all directions. After the uniform expansion, relation d_{364} would obtain between W and Y but not between X and Y; and relation d_{593} would obtain between X and Z after the expansion though it obtained between *no* spatial entities before the expansion. This example illustrates that, if Spatial Directionalism is correct, then UE-2 is false: If there exist spatial directional relations, then it is false that a non-zero-dimensional spatial entity would involve the same spatial relations after uniform expansion that it involved prior to its expansion. And, this objection to the Uniform Expansion Argument *does* leave open the question of whether the Spatial Directionalist should side with Spatial Substantivalism or Spatial Relationalism.

THE ABSOLUTE MOTION ARGUMENT FOR SS

Tim Maudlin has noted the significance of space and motion with respect to physics:

> The importance of an account of space in the formulation of physics cannot be overstated. If physics is first and foremost about motion, and motion is change of place, then (it seems) there must be *places* that material objects can successively occupy. An object rests when it occupies the same place over time, like Aristotle's stone at the center of the universe. It is tempting to say that without some sort of space in which things move, physics cannot even get off the ground.[19]

To explain certain physical phenomena involving certain objects, Newton postulated the existence of certain forces; and, to explain the origin of these forces, Newton argued that the *absolute* motion of the objects generates the forces. Then, to explain what *absolute* motion involves, Newton postulated the existence of *absolute* (i.e., substantivalist) *space*, claiming that an object moves *absolutely* when it occupies successively different regions of substantivalist *space*. Consider Newton's two examples of physical phenomena in need of an explanation.

The first example involves a spinning bucket of water:

> If a vessel hung by a long cord, is so often turned about that the cord is strongly twisted, then filled with water, and held at rest together with the water; thereupon, by the sudden action of another force, it is whirled about the contrary way, and while the cord is untwisting itself, the vessel continues for some time in this motion; the surface of the water will at first be plain, as before the vessel began to move, but after that, the vessel, by gradually communicating its motion to the

water, will make it begin sensibly to revolve, and recede by little and little from the middle, and ascend to the sides of the vessel, forming itself into a concave figure ... and the swifter the motion becomes, the higher will the water rise.[20]

When a bucket of water suspended from a twisted rope is held at rest, the water's surface is flat; when the bucket is released and allowed to spin, the water's surface becomes concave as the water climbs the wall of the bucket.

The second example involves something like a bola spinning through the air—two spheres revolving rapidly around the center of a cord that connects them:

> [I]f two globes, kept at a given distance one from the other by means of a cord that connects them, were revolved about their common centre of gravity, we might, from the tension of the cord, discover the endeavor of the globes to recede from the axis of their motion ... even in an immense vacuum, where there was nothing external or sensible with which the globes could be compared. But now, if in that space some remote bodies were placed that kept always a given position one to another, as the fixed stars do in our regions, we could not indeed determine from the relative translation of the globes among those bodies, whether the motion did belong to the globes or to the bodies. But if we observed the cord, and found that its tension was that very tension which the motions of the globes required, we might conclude the motion to be in the globes, and the bodies to be at rest.[21]

Newton considers two scenarios in which the cord connecting two spheres would exhibit tension. In a world in which the revolving spheres and cord (and their parts) were the only spatial entities, the cord *would* exhibit tension even though the revolving spheres would not be undergoing *relative* motion—even though there would exist no other spatial entities relative to which the spheres would be moving. In another world in which the cord-connected spheres exist and are surrounded by other spatial entities, the cord would exhibit tension even though it would be in principle impossible to determine whether (a) the cord-connected spheres are revolving relative to the world's other (fixed) spatial entities or (b) the other spatial entities are revolving around, and thereby moving relative to, the nonrevolving cord-connected spheres. Newton observes that, in both scenarios, the cord would exhibit tension when there is no determinate *relative* motion of the cord-connected spheres.

Newton concludes that, by positing substantivalist *space*, one can best explain the deformation of the water's surface and the cord's tension. Maudlin succinctly summarizes Newton's reasoning:

> Newton's bucket experiment, for all its simplicity, remains one of the most powerful and compelling experiments in the history of physics. The behavior of

the water in the bucket, or the tension in the cord connecting the globes, is an observable fact that requires an explanation. The natural explanation occurs to us: the water ascends the sides of the bucket and cord displays tension because the system is *spinning*. But spinning is a sort of motion, so we must ask: spinning *with respect to what*? The relevant motion is not motion with respect to the immediate surroundings. And if we accept the thought experiment with the globes in an otherwise empty space, the relevant motion cannot be motion with respect to any material body. Newton concludes that the motion must be motion with respect to absolute space: the spinning bodies successively occupy different locations in space itself. In this way, absolute motions are connected to forces and hence to observable effects.[22]

In short, Newton claims that, in a world that contains nothing other than the spinning bucket or the revolving globes, the deformation of the water's surface or the cord's tension would be caused by forces that are generated by the actual (i.e., absolute) motion of the bucket or the globes; and actual (i.e., absolute) motion would involve the bucket's parts or the globes occupying different regions of substantivalist (absolute) *space* at different times. After all, in such a world, motion would be impossible without absolute space: Per Leibniz, without absolute space, motion could involve nothing more than an object's changing the spatial relations that it bears to other objects; and in a world that contains nothing other than the spinning bucket or the revolving globes, there would exist no other objects relative to which the bucket or globes could change spatial relations.[23] But without motion, one could *not* explain the deformation of the water's surface or the cord's tension by *motion*-caused forces. Thus, Newton argued, explaining the deformation and the tension by *motion*-caused forces requires commitment to *absolute* motion—motion that involves an entity's successively changing its location in substantivalist *space*.

Below is a more formal explication of Newton's Absolute Motion argument for SS:[24]

AM 1. There is a possible world *b* in which there exists nothing but a rapidly spinning bucket half-filled with water with a concave surface, and there is a possible world *g* in which there exists nothing but two cord-connected globes revolving around the center of the cord that itself exhibits tension.
 2. If *b* and *g* are possible worlds, then it is reasonable to believe that, in *b* and *g*, there exist forces that contribute causally to the deformation of the water's surface and to the cord's tension and that these forces are caused by the motion of the bucket and the globes.
 3. Therefore, it is reasonable to believe that, in *b* and *g*, there exist forces that contribute causally to the deformation of the water's surface and to

the cord's tension and that these forces are caused by the motion of the bucket and the globes. (from AM-1, AM-2)

4. If it is reasonable to believe that, in *b* and *g*, there exist forces that contribute causally to the deformation of the water's surface and to the cord's tension and that these forces are caused by the motion of the bucket and the globes, then it is reasonable to believe that the forces in *b* and *g* would be caused by the relative motion or by the absolute motion of the bucket and globes.

5. Therefore, it is reasonable to believe that the forces in *b* and *g* would be caused by the relative motion or by the absolute motion of the bucket and globes. (from AM-3, AM-4)

6. If the forces in *b* and *g* would be caused by the relative motion of the bucket and globes, then (i) in *b*, there would exist a spatial entity *x* discrete from the bucket of water such that the bucket would move relative to *x* (i.e., the spatial relations that obtain between the bucket and *x* would change) and (ii) in *g*, there would exist a spatial entity *y* discrete from the cord-connected globes such that the cord-connected globes move relative to *y* (i.e., the spatial relations that obtain between the cord-connected globes and *y* would change).

7. It is false that, in *b*, there would exist a spatial entity *x* discrete from the bucket of water, and it is false that, in *g*, there would exist a spatial entity *y* discrete from the cord-connected globes.

8. Therefore, it is false that the forces in *b* and *g* would be caused by the relative motion of the bucket and globes. (from AM-6, AM-7)

9. Therefore, it is reasonable to believe that the forces in *b* and *g* would be caused by the absolute motion of the bucket and globes. (from AM-5, AM-8)

10. If the forces in *b* and *g* would be caused by the absolute motion of the bucket and globes, then there would exist in *b* and *g* regions of substantivalist *space* such that the bucket and cord-connected globes undergo absolute motion, occupying different regions of *space* at different times.

11. Therefore, it is reasonable to believe that there would exist in *b* and *g* regions of substantivalist *space* such that the bucket and cord-connected globes undergo absolute motion, occupying different regions of *space* at different times. (from AM-9, AM-10)

12. If it is reasonable to believe that there would exist in *b* and *g* regions of substantivalist *space* such that the bucket and cord-connected globes undergo absolute motion, occupying different regions of *space* at different times, then it would also be reasonable to believe that there actually exists substantivalist *space* relative to which actual spatial entities move absolutely.

13. Therefore, it is reasonable to believe that there actually exists substantivalist *space* relative to which actual spatial entities move absolutely. (from AM-11, AM-12)

In his fourth letter to Leibniz, Clarke endorses Newton's Absolute Motion Argument for substantivalist *space* and presses Leibniz for a response, cautioning him not to dismiss Newton's argument by flatly endorsing Spatial Relationalism instead:

> The motion of the ship, though the man [shut up in the cabin] perceives it not, is a real different state, and has real different effects; and, upon a sudden stop, it would have other real effects; and so likewise would an indiscernible motion of the universe. To this argument, no answer has ever been given. It is largely insisted on by Sir Isaac Newton in his *Mathematical Principles* (Defnit.8.) where . . . he shows the difference between real motion, or a body's being carried from one part of space to another; and relative motion, which is merely a change of the order or situation of bodies with respect to each other. This argument is a mathematical one; showing, from real effects, that there may be real motion where there is none relative; and relative motion, where there is none real: and is not to be answered, by barely asserting the contrary.[25]

Appearing to take seriously Clarke's admonishment that he should not simply endorse Spatial Relationalism and stubbornly embrace its implication that Newton's argument is flawed, Leibniz instead grants (in his fifth paper to Clarke) that there *is* a distinction between *absolute* and *relative* motion and that the distinction can be preserved *without* commitment to substantivalist *space*:

> I grant there is a difference between an absolute true motion of a body, and a mere relative change of its situation with respect to another body. For when the immediate cause of the change is in the body, that body is truly in motion; and then the situation of other bodies, with respect to it, will be changed consequently, though the cause of that change be not in them. 'Tis true that exactly speaking there is not any one body, that is perfectly and entirely at rest; but we frame an abstract notion of rest, by considering the thing mathematically. Thus have I left nothing unanswered, of what has been alleged for the absolute reality of space.[26]

Presumably, one should construe Leibniz's response as an objection to AM-10 to the effect that absolute motion does *not* imply the existence of substantivalist *space*. Leibniz's response, however, is, at best, puzzling.

First, what *could* count as an "immediate cause" of a change that involves a particular object? And, what could be the difference between an "immediate cause" and a "*non*-immediate cause" of an object's change? Without plausible answers to such questions, one cannot determine whether the "absolute" spinning of the bucket or the "absolute" revolution of the globes can be explained in terms of "immediate causes" and without an appeal to substantivalist *space*.

Second, why should one suppose that the "immediate cause" of an object's change would be the sort of thing that would be "in" the object? Why *would* the "immediate causes" of the bucket's "absolute" spinning and the globes' "absolute" revolution be "in" (rather than "outside"?) the bucket and the globes?

Third, if one grants that an object undergoes a change that is caused by an "immediate cause" that *is* "in" the object, it does not obviously follow that the object is thereby in "absolute true motion." In the end, then, Leibniz's response to the Absolute Motion Argument for SS remains entirely unclear: It is unclear how the "absolute true motion" of the spinning bucket and the revolving globes can be explained in terms of "mere relative change" when, in worlds *b* and *g*, there would exist nothing relative to which the bucket and cord-connected globes could undergo "mere relative change."[27]

Earlier in his fifth paper to Clarke, Leibniz suggests a different relationalist response to the Absolute Motion Argument, but it *is* the response that Clarke cautions against:

> I don't grant that every finite is moveable. . . . What is moveable, must be capable of changing its situation with respect to something else, and to be in a new state discernible from the first: otherwise the change is but a fiction.[28]

Leibniz's claim is that only *relative* motion is possible—that an object cannot possibly move unless there exists at least one other (nonoverlapping) object relative to which it can change spatial relations. Berkeley also endorses this view regarding the nature of motion:

> [I]t doth not appear to me, that there can be any motion other than *relative* : so that to conceive motion, there must be at least conceived two bodies, whereof the distance or position in regard to each other is varied. Hence, if there was one only body in being, it could not possibly be moved.[29]

> As to what is said of the centrifugal force, that it does not at all belong to circular relative motion : I do not see how this follows from the experiment which is brought to prove it. . . . For the water in the vessel, at that time wherein it is said to have the greatest relative circular motion, hath, I think, no motion at all.[30]

Leibniz and Berkeley's claim that no spatial entity can possibly move in a world in which there exists no other (nonoverlapping) spatial entity can be construed as an objection to AM-1: There is *no* possible world in which there exists nothing other than a spinning bucket or revolving cord-connected globes.

This objection to AM-1 indeed violates Clarke's warning to Leibniz that one should not object to substantivalist *space* "by barely asserting the contrary"—by stubbornly affirming SR and its implication that absolute motion is impossible. Arguably, the Spatial Relationalist *does* owe the Spatial Substantivalist more than a simple rejection of AM-1 on grounds that it is inconsistent with SR.

Regardless of whether one is a Spatial Relationalist, there are other reasons to be skeptical about AM-1. *Is* there a possible world in which there exists nothing but a bucket of water connected to the end of a twisted cord? *Is* there a possible world in which there exists nothing but the cord-connected-globes? *Would* the actual laws of nature or *any* laws of nature allow for the existence of a stark world in which a solitary object and its parts are the only existing spatial entities? In such a stark world, how, exactly, is it possible that the matter would have come into being or would have existed without beginning? How, exactly, is it possible that some of the matter in such a stark world would have become the sort of matter that could compose a bucket or a cord while other matter would have become the sort of matter that could compose water or a globe? How, exactly, is it possible that all the matter of a certain type would compose a bucket or a cord (rather than a platter or a handkerchief) while all the matter of another type would compose iron globes or a glob of mercury (rather than horseshoes or a hunk of clay)? Without plausible answers to these questions, AM-1 will remain, at best, controversial.

Though there are reasons to be skeptical about AM-1, there is nonetheless something to be said for Newton's insistence that an adequate explanation for the deformation of the water's surface and the cord's tension should at least allow for the possibilities that these phenomena would occur even in the stark worlds in which there exists no (nonoverlapping) objects other than, respectively, the spinning bucket and revolving cord-connected globes. But regardless of what one concludes about the truth value of AM-1, one can nonetheless resist the Absolute Motion Argument by affirming Spatial Directionalism and objecting to AM-10: One can use spatial *directional* relations to formulate a relationalist account of absolute motion that does *not* involve substantivalist *space*. And, this relationalist account of absolute motion *is* consistent with the Newtonian conviction that there exists a possible world that contains nothing but a *spinning* bucket of water or *revolving* cord-connected globes.

To understand the Spatial Directionalist's account of absolute motion, imagine a *directioned* world that contains no spatial entity but a spinning bucket of water and its parts. The Spatial Directionalist would claim that the bucket undergoes absolute motion in the sense that it involves different spatial *directional* relations as it spins. Imagine that the bucket (represented in figure 4.4) is composed of bucket-halves AA and BB and that, at time T, (a) AA (but not BB) involves spatial *directional* relation d_{238} and (b) BB bears spatial directional relation d_{981} to AA. As the bucket spins counterclockwise one half-turn between T and T*, there *is* a change of spatial (directional) relations: At T*, BB (but not AA) involves d_{238} and AA bears spatial directional relation d_{981} to BB (not the other way around). By claiming that the spinning bucket successively exhibits different spatial *directional* relations, the Spatial Directionalist can agree with Leibniz that motion is impossible without a spatial entity's changing the spatial relations that it bears to other entities; *and* the Spatial Directionalist can also agree with Newton that the deformation of the water's surface is caused by forces generated by the *absolute* motion of the bucket.[31]

To explain the cord's tension in a world that includes no spatial entity other than *revolving* cord-connected-globes, the Spatial Directionalist could claim that, as the globes revolve, each globe would successively involve a different set of spatial directional relations. For example, the spatial directional

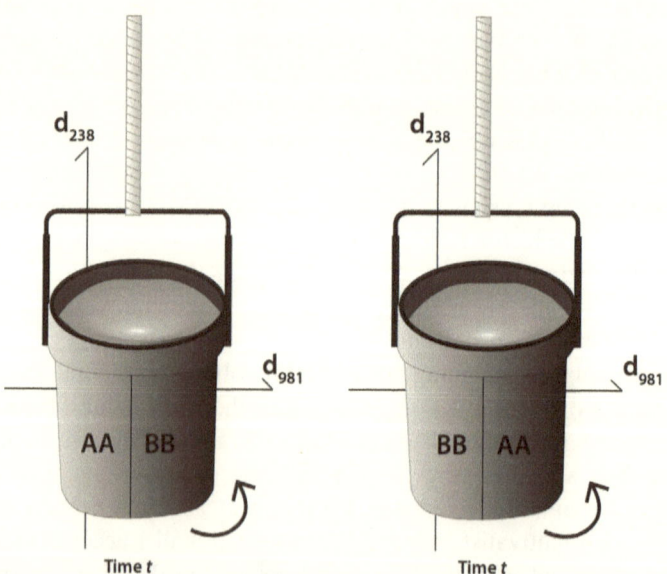

Figure 4.4 Absolute Motion per Spatial Directionalism. *Source*: C. Frantom, J. Rich: White Roche LLC.

relations that globe #1 would bear to globe #2 at a given time would be other than the relations that #1 bears to #2 after the cord-connected-globes undergo a quarter-revolution or a half-revolution or a three-quarter-revolution. (Of course, the spatial directional relations that one cord-half would bear to the other cord-half would also change from time to time as the cord-connected-globes revolve.)

Thus, by appealing to spatial directional relations, the Spatial Directionalist can preserve the robust distinction between absolute and relative motion without positing substantivalist space. And this would allow the Spatial Directionalist to deny AM-10 and thereby reject Newton's Absolute Motion argument for SS.[32]

AN OBJECTION TO SPATIAL DIRECTIONAL RELATIONS

The Spatial Substantivalist should agree with the Spatial Directionalist that the Leibnizian objections to the standard arguments for SS are implausible; but some would argue that the Spatial Directionalist's objections to the standard arguments are at least as implausible:

> Only by positing the existence of (absolute) spatial directional relations can the Spatial Directionalist preserves the plausible substantivalist intuitions involving uniform motion, the inverted world, the uniform expansion of objects, and absolute motion. But there appears to be no criterion of identity with respect to spatial directional relations—no criterion that could explain why spatial directional relation d is or is not identical with spatial directional relation d'. With no criterion of identity, however, one should deny the existence of spatial directional relations.

Below is a more formal version of this Identity Criterion Objection to Spatial Directionalism:

IC 1. Spatial Directionalism is correct only if spatial directional relations exist.
   2. Spatial directional relations exist only if there exists a criterion of identity for spatial directional relations.
   3. There is no criterion of identity for spatial directional relations.
   4. Therefore, spatial directional relations do not exist. (from IC-2, IC-3)
   5. Therefore, SD is not correct. (from IC-1, IC-4)

Regarding IC, there are several replies available to the Spatial Directionalist.

First, with respect to IC-3, one should demand no more (or less) of the Spatial Directionalist than one demands of the *location* substantivalist who insists that points of *space* (or *spacetime*) exist. So, if it is fair to demand criteria of identity for spatial directional relations, it is also fair to demand criteria of identity for points of *space*: With respect to a point of *space* (or *spacetime*) p, what criterion of identity could explain whether p is or is not identical with point of space p'? The substantivalist should not claim that spatial point p is identical with p', if and only if, p and p' are located in exactly the same place: Spatial points cannot possibly be *located* in a *place* given that points *are* places where *other* things (i.e., spatial entities) are located. And, one should not formulate the criterion of identity for *spatial* (or spacetime) points in terms of sameness of color, mass, charge, or spin: Spatial points essentially *lack* such characteristics. (In Leibniz's words, cited earlier, "without the things placed in it, one point of space does not absolutely differ in any respect whatsoever from another point of space."[33])

One may be tempted to formulate the criterion of identity for spatial points in terms of spatial entities:

Spatial point p is identical with p', if and only if, for any time t, either nothing occupies p and p' or the spatial entity that occupies p also occupies p'.

By implying that spatial point p would be identical with p' if the *same* spatial entity occupies p and p', the above criterion of identity for spatial *points* presupposes a criterion of identity for spatial *entities*. But what *is* the criterion of identity for *spatial entities*? If one mereological simple occupies spatial point p and a different mereological simple occupies p', what criterion of identity would explain the difference between those two simples—simples that, one may argue, may not differ with respect to their intrinsic properties. One may be tempted by the view that the two simples would differ with respect to their *locations*—they would differ with respect to which points of space they occupy. This appeal to location is, for the substantivalist, unacceptable: The substantivalist who attempts to explain the difference between spatial points in terms of spatial entities and *then* appeals to spatial points to explain the difference between spatial entities would be subject to the charge of circularity.

Perhaps the substantivalist will insist that spatial points can be differentiated with respect to "excitation" or "field values" or some such. But would it be true that no two spatial points could possibly share the same degree of "excitation" or the same "field value"?[34] If not, then what criterion of identity could explain the difference between two spatial points that share the same degree of "excitation" or the same "field value"? Suppose that, ultimately, the substantivalist admits that there is no plausible criterion of identity for points of space (or spacetime), but is unwilling to abandon ontological commitment

Defending Spacelessness

to points of space given their significant theoretical benefit. Such a substantivalist could argue that one should sidestep the demand for a criterion of identity, insisting that the evidence for spatial points is sufficiently compelling to warrant the inference that either (a) spatial points indeed have a yet-unknown criterion of identity (e.g., perhaps each has its own "primitive *thisness*") *or* (b) there do exist some things (e.g., points of space or spacetime) that lack criteria of identity:

ICP 1. There is good evidence that points of *space* (or *spacetime*) exist.
     2. If points of *space* (or *spacetime*) exist, then either (a) there *is* a (known or unknown) criterion of identity for such points *or* (b) there do exist some things (e.g., points of *space* or *spacetime*) that lack criteria of identity.
     3. Therefore, there is good evidence that either (a) there *is* a (known or unknown) criterion of identity for points of *space* (or *spacetime*) *or* (b) there do exist some things (e.g., points of *space* or *spacetime*) that lack criteria of identity.

The substantivalist who defends ICP could simply leave open the question of whether there is or cannot be a criterion of identity for *spatial* (or *spacetime*) points.

A similar move is available to the Spatial Directionalist with respect to spatial directional relations:

ICR 1. There is good evidence that spatial directional relations exist.
     2. If spatial directional relations exist, then either (a) there *is* a (known or unknown) criterion of identity for such relations *or* (b) there do exist some things (e.g., spatial directional relations) that lack criteria of identity.
     3. Therefore, there is good evidence that either (a) there *is* a (known or unknown) criterion of identity for spatial directional relations (i.e., IC-3 is false) *or* (b) there do exist some things (e.g., spatial directional relations) that lack criteria of identity (i.e., IC-2 is false).

A Spatial Directionalist, then, could emphasize the significant philosophical benefit of endorsing spatial directional relations and then leave open the question of whether there is or could be a criterion of identity for such relations (or for *any* relations).

Other Spatial Directionalists may resist sidestepping the demand for a criterion of identity and try to meet the critic's challenge. For example, a directionalist may be tempted by the following criterion of identity for spatial directional relations:

Spatial directional relation d is identical with d', if and only if, d and d' are necessarily such that, for any time t, d obtains between two spatial entities, if and only if, d' also obtains between those two spatial entities.

This criterion is inadequate: A spatial directional relation and the opposite relation are such that they both obtain between exactly the same spatial entities; and although such relations would thereby satisfy the criterion's conditions for identity, the relations are *not* identical with one another.

The Spatial Directionalist may suggest instead that the criterion of identity for spatial directional relations should be cast in terms of the spatial relations that one spatial entity *bears* to another:

> Spatial directional relation d is identical with d', if and only if, d and d' are necessarily such that any spatial entity that bears d to a second spatial entity also bears d' to that second entity.

This criterion is also inadequate. Consider the relation *bears a spatial directional relation to*. Any spatial entity s_1 that bears, say, specific spatial directional relation d_{37} to a second entity s_2 also bears *bears a spatial directional relation to* to s_2. Although d_{37} and *bears a spatial directional relation to* satisfy the criterion above, d_{37} is *not* identical with *bears a spatial directional relation to*: After all, it is possible that *bears a spatial directional relation to* obtains when d_{37} does not. For example, spatial entities s_1 and s_2 could move in such a way that s_1 ceases to bear d_{37} to entity s_2 but continues to bear *bears a spatial directional relation to* to s_2.

At the very least, then, the Spatial Directionalist should appeal to the concept of *implication*:

> D4.1 Spatial directional relation d implies d' =Df d and d' are necessarily such that, for any x and y, if x bears d to y, then x also bears d' to y.

Spatial directional relation d implies d' when anything that bears d to something also bears d' to that something. Consider, then, a criterion of identity cast in terms of mutual implication:

> Spatial directional relation d is identical with d', if and only if, d implies d' and d' implies d.

This criterion alone implies, as it should, that d_{37} is *not* identical with *bears a spatial directional relation to*. Although *bears d_{37} to* implies *bears a spatial directional relation to*, *bears a spatial directional relation to* does not imply *bears d_{37} to*: A spatial entity that bears *bears a spatial directional relation to* a second spatial entity may fail to bear *bears d_{37} to* that second entity. Thus, *bears d_{37} to* and *bears a spatial directional relation to* would thereby fail to satisfy the above "mutual implication" criterion of identity.

Though the "mutual implication" criterion of identity may be tempting, some may argue that implication alone is not sufficient for identity. Imagine a world in which, at a given time t, there exists nothing but a person R and two mereological simples s_1 and s_2 such that s_1 bears d_{37} to s_2, and s_2 bears opposite relation d_{37*} to s_1. Suppose that R knows that the only spatial directional relation that s_1 bears to s_2 is the unique spatial directional relation d_{37}. Suppose as well that, although R does infer that s_2 thereby bears *bears the spatial directional relation that is the opposite of d_{37} to* to s_1, R does not know *which* specific spatial directional relation *is* the opposite of d_{37}: For all R knows, d_{37}'s opposite is d_{37*} or d_{84*} or d_{194*}. Therefore, although R believes that s_2 bears *bears the spatial directional relation that is the opposite of d_{37} to* to s_1, R does not believe that s_2 bears d_{37*} to s_1. Thus, one may plausibly object, the "mutual implication" criterion of identity given that d_{37*} is *not* identical with *bears the spatial directional relation that is the opposite of d_{37} to* even though these *two* properties imply one another and would thereby satisfy the "mutual implication" criterion of identity.

Instead of appealing to the concept of implication alone, the Spatial Directionalist could also appeal to the concept *involvement*—a concept that Chisholm introduces when addressing the relations that obtain between properties:[35]

> D4.2 Spatial directional relation d involves d' =Df d is necessarily such that whoever conceives it also conceives d'.

Although *bears the spatial directional relation that is the opposite of d_{37} to* and d_{37*} do *imply* one another, they do not *involve* one another. The person R who coexists with s_1 and s_2 believes that s_2 bears the spatial directional relation that is the opposite of d_{37} to s_1 and thereby conceives relation d_{37}; but it is possible that R is entirely unaware that d_{37*} exists and thereby would thereby not conceive of d_{37*} even though d_{37*} implies d_{37}.

The concept of involvement allows the Spatial Directionalist to formulate a criterion of identity for spatial directional relations in terms of both mutual implication and mutual involvement:

> Spatial directional relation d is identical with d', if and only if, d and d' are necessarily such that (i) d implies d', and d' implies d, and (ii) d involves d' and d' involves d.

This "mutual involvement" criterion implies that d_{37*} is *not* identical with *bears the spatial directional relation that is the opposite of d_{37} to*: Even though these *two* properties imply one another, they do *not* involve one another given that one may conceive of *bears the spatial directional relation that is the opposite of d_{37} to* without conceiving of d_{84}.

Summary regarding the Identity Criterion Objection to Spatial Directional Relations. To defend spatial directional relations against the IC objection, several replies are available to Spatial Directionalists: (a) Some Spatial Directionalists may argue that there do exist *some* things that lack criteria of identity and that spatial directional relations are among these things; (b) other Spatial Directionalists may argue that spatial directional relations exist and *do* have a criterion of identity that remains to us unknown; and (c) still other Spatial Directionalists may argue that spatial directional relations exist and that the "mutual implication/involvement" criterion *is* the criterion of identity for spatial directional relations.

ASSESSING THE ROAD TO NO*WHERE*

A Spatial Directionalist could indeed remain a skeptic regarding the existence of substantivalist *space*; but a Spatial Directionalist who is committed to a leaner ontology or who finds the concept of substantivalist *space* hopelessly mysterious or problematic could instead side with the Directionalist Theory of Space and thereby with (non-Leibnizian) Spatial Relationalism. Consider a modest defense of DTS that is available to such a directionalist.

First, the Spatial Directionalist could note advantages of Spatial Directionalism itself—the advantages of positing the existence of spatial *directional* relations. By embracing spatial directional relations, one can thereby distinguish spatial entities from other entities: Spatial entities—entities that *"have locations"*—are those things that possibly bear spatial directional relations to other things. And, as addressed in chapter 3, not only is the existence of spatial directional relations consistent with the existence of more than three spatial dimensions (i.e., with Hyperspace), but embracing spatial directional relations would allow one to explicate the concept of spatial dimensionality in terms of such relations.

Second, also addressed in chapter 3, the Spatial Directionalist can note that plausible claims that purportedly imply that substantivalist *space* exists *are* reducible to claims involving spatial directional relations that clearly do *not* imply that *space* exists. For example, claims that seem to imply that there exists a *place* that a spatial entity occupies can be reformulated in terms of the spatial directional relations that the entity *involves*. And, as addressed earlier in the present chapter, claims that imply that absolute motion involves moving from one region of *space* to another can be reformulated in terms of a spatial entity's changing the spatial directional relations that it bears to others.

Third, as explained in the first four sections of this chapter, the Spatial Directionalist can appeal to spatial directional relations both to reject the classic arguments for Spatial Substantivalism and to preserve the substantivalist intuitions that underlie those arguments—plausible substantivalist intuitions regarding uniform motion, spatial orientation, uniform expansion, and absolute motion.

Finally, citing Newton's caveat that one should "*admit no more causes of natural things than such as are both true and sufficient to explain their appearances*,"[36] the Spatial Directionalist can note that, *ceteris paribus*, ontological simplicity favors DTS over SS.

The cautious Spatial Directionalist can weave these four considerations into a modest defense of DTS: "Given that there is no compelling argument that substantivalist *space* exists and given that plausible *space*-implying claims are reducible to non-*space*-implying claims cast in terms of spatial directional relations, it is reasonable to reject SS and side with the ontologically leaner DTS, affirming that spatial directional relations exist and that substantivalist *space* does not exist." Below is a formulation of this defense of DTS:

ADTS 1. If (i) one lacks a plausible argument on behalf of the claim that X exists and (ii) one can reformulate plausible claims that purportedly imply that X exists into claims that clearly do not imply that X exists, then one has good reason to believe that X does *not* exist.
   2. The Spatial Directionalist lacks a plausible argument on behalf of the claim that substantivalist space exists.
   3. The Spatial Directionalist can reformulate plausible claims that purportedly imply that substantivalist space exists into claims involving spatial directional relations that clearly do not imply that substantivalist space exists.
   4. Therefore, the Spatial Directionalist has good reason to believe that substantivalist space does not exist. (from ADTS-1, ADTS-2, ADTS-3)

The Spatial Directionalist who affirms ADTS-3 and ADTS-4 would thereby affirm DTS; and given that spatial *directional* relations *are* spatial relations, the Spatial Directionalist who affirms DTS would thereby affirm Spatial Relationalism as well.

To strengthen the case for DTS (and thereby for SR), the defender of DTS should couple ADTS with positive reasons for rejecting the existence of *space*. For example, some DTS defenders may develop Berkeley's objection (mentioned above in chapter 3) that the concept of space may be utterly mysterious—that it is "impossible" to "frame an idea of *pure space*."

One could also strengthen the case for DTS by demonstrating that DTS and its commitment to spatial directional relations can provide plausible, simpler resolutions to still other philosophical problems. This abductive case for DTS will be developed in the next two chapters. In chapter 5, I invoke spatial directional relations to develop a noncircular reformulation of my original answer to the Special Composition Question—an answer that preserves the commonsense views that organisms, rocks, planets, and stars *are* among the spatial entities that exist and that some spatial entities (e.g., molecules) are "scattered objects." In chapter 6, I use DTS to develop reductivist theories of boundaries and holes—theories that would *not* commit one to the existence of strange, "dependent," non-three-dimensional parts of objects *or* to the existence of "immaterial" or "negative" entities that are parts of objects. If reductivist theories of boundaries and holes formulated in terms of DTS do allow one to skirt ontological commitment to bizarre *dependent particulars*, then this would be yet another mark in favor of DTS.

NOTES

1. For a succinct summary of Leibniz's defense of Spatial Relationalism, see Brian Greene, *The Fabric of the Cosmos*, p. 30.
2. C.III.4, *The Leibniz-Clarke Correspondence*, p. 32.
3. See Shamik Dasgupta's formulation and discussion of Leibniz's argument in "Substantivalism vs Relationalism About Space in Classical Physics," 606–20. See also Maudlin, "Buckets of Water and Waves of Space," pp. 188–92. The possibility that all things move uniformly is what Maudlin calls the *"kinematic Leibniz shift."*
4. L.V.47, *The Leibniz-Clarke Correspondence*, pp. 69–70.
5. L.V.29, *The Leibniz-Clarke Correspondence*, p. 63. See L.IV.13; p. 38: "To say that God can cause the whole universe to move forward in a right line . . . without making otherwise any alteration in it; is another chimerical supposition." See also L.V.52.
6. See Dasgupta, "Substantivalism vs Relationalism About Space in Classical Physics," 617–18, 621. With respect to arguments involving uniform motion, Dasgupta entertains briefly "the qualitativist view of space-time"—the non-substantivalist view that "the fundamental facts about space-time are purely qualitative" insofar as "they would just describe a patchwork of purely qualitative geometric relations." I am not myself competent to determine whether the spacetime qualitativist could use a version of the Spatial Directionalist's objection to UM to craft a plausible qualitativist reply to the Uniform Motion Argument.
7. Brentano, "On *Ens Rationis*," p. 363.
8. C.II.1, *The Leibniz-Clarke Correspondence*, pp. 20–21.
9. Dean Rickles addresses the inverted-world defense of SS as well as the Leibnizian reply, casting the defense in terms of handedness and the problem of incongruent counterparts. See Rickles' "A Handy Argument for the Substantivalist" (Section 4.2) in his *The Philosophy of Physics* (Cambridge, UK, and Malden, MA: Polity Press, 2016), pp. 55–61. See also Maudlin, "Buckets of Water and Waves of Space," pp. 188–92. That the inverted version of the actual world could have existed instead is what Maudlin calls the *"static Leibniz shift."*
10. Leibniz, §32, *Monadology*, p. 153: "[N]o fact can be true or existing and no statement truthful without a sufficient reason for its being so and not different."

11. L.III.5, *The Leibniz-Clarke Correspondence*, p. 26.

12. See also Friedman, *Foundations of Space-Time Theories*, pp. 218–19; Friedman offers a succinct interpretation of Leibniz's argument "freed of theology," noting that the argument is also applicable not only to substantivalist space but to substantivalist *spacetime*.

13. C.III.2, *The Leibniz-Clarke Correspondence*, pp. 30–31. See also C.III.6, pp. 32–33: "But when two ways of acting are equally and alike good, . . . to affirm in such case, that God cannot act at all . . . because he can have no external reason to move him to act one way rather than the other, seems to be a denying God to have in himself any original principle or power of beginning to act, but that he must needs (as it were mechanically) be always determined by things extrinsic." Cf. Peter van Inwagen, 4th ed., *Metaphysics* (Boulder, CO: Westview Press, 2015), pp. 164–67. van Inwagen argues that the Principle of Sufficient Reason implies counterintuitively that all truths are necessary truths.

14. C.IV.18, *The Leibniz-Clarke Correspondence*, pp. 49–50.

15. C.V.21–25, *The Leibniz-Clarke Correspondence*, pp. 99–100. See also C.V.1–20, p. 98: "To affirm therefore, . . . that supposing two different ways of placing certain particles of matter were equally good and reasonable, God could neither wisely nor possibly place them in either of those ways, for want of a sufficient weight to determine him which way he should choose; is making God not an active, but a passive being: which is, not to be a God, or governor, at all."

16. Brentano, "What we can learn about space and time from the conflicting errors of the philosophers" (dictated February 23, 1917), *Space, Time and the Continuum*, p. 172.

17. Henri Poincaré, *Science and Method*, trans. Francis Maitland (London, Edinburgh, Dublin, and New York: Thomas Nelson and Sons, 1914), p. 94.

18. Poincaré, *Science and Method*, p. 95.

19. Tim Maudlin, *Philosophy of Physics: Space and Time* (Princeton, NJ, and Oxford: Princeton Foundations of Contemporary Philosophy, an imprint of Princeton University Press, 2012), p. 4.

20. Def.8, Schol.IV, *Newton's Mathematical Principles*, p. 10.

21. Def.8, Schol.IV, *Newton's Mathematical Principles*, p. 12.

22. Maudlin, *Philosophy of Physics*, p. 23. See also Brentano's summary of Newton's defense; "What we can learn about space and time," p. 162. See also Julian Barbour, *The End of Time: The Next Revolution in Physics* (Oxford and New York: Oxford University Press, 1999), pp. 62–67.

23. Cf. Berkeley, 115, *A Treatise Concerning the Principles of Human Knowledge*, p. 92. Berkeley claims that a body can be in motion only if both "it change its distance . . . with regard to some other body : and . . . the force or action occasioning that change be applied to it"; if either of these conditions fails to obtain, concludes Berkeley, "I do not think that . . . a body can be said to be in motion."

24. See Dasgupta, "Substantivalism vs Relationalism About Space in Classical Physics," 603–05 and Maudlin, "Buckets of Water and Waves of Space," 186. Dasgupta and Maudlin offer formulations of this "bucket" argument in terms of an actual experiment and in terms of a thought experiment.

25. C.IV.13, *The Leibniz-Clarke Correspondence*, p. 48.

26. L.V.53, *The Leibniz-Clarke Correspondence*, p. 74.

27. In a fair assessment of Leibniz's admission that "'there is a difference between an absolute true motion of a body, and a mere relative change of its situation with respect to another body,'" Brian Greene notes that Leibniz's concession "was not a capitulation to Newton's absolute space, but it was a strong blow to the firm relationalist position." See Greene's, *The Fabric of the Cosmos*, p. 31.

28. L.V.31, *The Leibniz-Clarke Correspondence*, p. 64.

29. Berkeley, 112, *A Treatise Concerning the Principles of Human Knowledge*, p. 91. See also 450a, *Philosophical Commentaries*, Notebook A, p. 56: "Motion distinct from ye thing moved is not Conceivable." See also 876, p. 103: "If there were only one Ball in the World it Could not be moved."

30. 114, *A Treatise Concerning the Principles of Human Knowledge*, p. 92.

31. Emory Kimbrough brought to my attention Julian Barbour's discussion of Newton's mechanics vis-à-vis substantivalist *space* and time. Barbour explains why Poincaré concluded that "a [relationalist] mechanics that uses only relative quantities, as Mach advocated, cannot get off the ground." (Barbour addresses absolute motion vis-à-vis substantivalist *spacetime* in chapter 7.) See Barbour, *The End of Time*, see chs. 1, 3, 4, and 5. In chapter 7, Barbour writes that he and Bruno Bertotti together "laid the foundations of a genuine Machian theory of the universe" that "showed that a mechanics of the complete universe containing only relative quantities and no extra Newtonian [substantivalist] framework could be constructed" (p. 114). I am not competent to address whether the absolute motion that Spatial Directionalism allows could prove helpful to Barbour, Bertotti, and others who are sympathetic with a Machian relationalism.

32. See my later discussion of Ernst Mach's commitment to relationalism in chapter 7. Tracing Mach's view to George Berkeley's *De Motu* (1721), Emory Kimbrough reports that Mach believed that the inertia and centrifugal force that an object exhibits would be the result of the object's interacting with the sum of all the other mass in the universe. On Mach's view, then, in a universe in which there exists nothing but a spinning bucket or bola, there would exist no other mass with which the bucket or bola could interact and there would thereby be no inertial phenomena generated. And, without inertia and centrifugal force, the solitary bucket or bola would spin with no deformation of the water's surface or no tension on the bola's cord. Interestingly, though Mach's view regarding inertial phenomena was motivated by his commitment to relationalism, Mach's view is consistent with Spatial Substantivalism. Thus, Mach's view allows for the possibility that a solitary bucket or bola could spin in empty substantivalist space *without* deformation of the water's surface or without the cord's exhibiting tension!

33. L.III.5, *The Leibniz-Clarke Correspondence*, p. 26.

34. Cf. Theodore Sider's defense of the view that spacetime points are "truly bare particulars"; "'Bare Particulars,'" *Philosophical Perspectives* 20 (2006), 387–97.

35. See Chisholm, "Properties and States of Affairs," p. 143; Chisholm introduces the concept of involvement when considering various relations that can obtain among properties. See also Chisholm's discussion of the criterion of identity for properties

on pp. 144–45 and in his essay, "Identity Criteria for Properties," *The Harvard Review of Philosophy* 2 (1992), 14–46.

36. Bk.III, *Newton's Mathematical Principles*, p. 398. Cf. Aristotle, I.25.1, *Posterior Analytics*, trans. A. J. Jenkinson, *The Basic Works of Aristotle*, ed. Richard McKeon (New York: Random House, 1941): "We may assume the superiority *ceteris paribus* of the demonstration which derives from fewer postulates or hypotheses."

Chapter 5

The Special Composition Question Revisited

THE QUESTION AT STAKE

Most assume that there exist objects composed of various parts. To put the point more formally, most assume that there exist three-dimensional spatial entities composed of nonoverlapping (i.e., mereologically discrete) proper parts.[1] Most would assume, then, that a table is an object composed of a table-top fastened to table-legs and that a planet is a large object composed of many millions of atoms and subatomic particles. Most would also assume that there is no object composed of their left shoe and Saturn. In 1981, I raised, and offered an answer to, a fundamental question about objects and their parts: If a certain table-top and certain table-legs *do* compose a whole when one's left shoe and Saturn do not, then when is it that objects *do* compose a whole and when do they not? Put another way, if several discrete objects do objects exist, but do *not* compose a whole? And what then must one do to bring it about that the nonoverlapping objects do again compose a whole? This question of when it is objects compose a whole is now known as "The Special Composition Question" [SCQ].[2]

The answer that I defended (and will continue to defend) is *Conjoining*: Nonoverlapping spatial entities compose something, when and only when, those entities are *conjoined*—when nothing lies between them. I will reformulate the 1981 analysis of what conjoining involves in terms of spatial directional relations. And, I will also use these relations to formulate the implications of Conjoining for the existence of spatial entities that lack proper parts (e.g., mereological atoms or monads), for the existence of three-dimensional *materially solid* objects, and for the existence of "scattered objects" (i.e., spatial entities with proper parts "entirely separated" by "empty space"). After formulating Conjoining, I explain why the reformulation is not subject to van Inwagen's "circularity" objection. To close the chapter, I offer replies to several potential objections to Conjoining. Consider first, however, the plausible presuppositions that motivate my answer to the Special Composition Question.

PRESUPPOSITIONS

There exist spatial entities composed of parts. Presupposed without argument is the possibility that there exist spatial entities that *have* proper parts.[3] That is, I reject Mereological Nihilism, which is the extreme view that no two spatial entities compose a third:

> MN For any x and y, if x and y are nonoverlapping spatial entities, then at no time does there exist something composed of x and y.[4]

Those who endorse MN would claim that, not only can there exist no wall composed of bricks, but there can exist no brick composed of half-bricks or quarter-bricks and no brick composed of brick molecules. Thus, the Mereological Nihilist's radical response to SCQ is that "*No* two spatial entities compose another!" I presuppose, then, that the Mereological Nihilist's answer to SCQ is mistaken and that one should search for a theory of composition that disallows MN's counterintuitive implication that there exist *no* spatial entities that are stones, dogs, trees, mountains, automobiles, or planets.

There exist objects that do not themselves compose an object. Just as one should resist the extreme view that there exist no spatial entities composed of parts, one should also resist Universalism—the extreme view that *any* two spatial entities compose a third:[5]

> UN For any two nonoverlapping spatial entities, x and y, there is a spatial entity that is composed of x and y at any time that x and y both exist.[6]

The Universalist's radical answer to SCQ is that "For *any* two spatial entities, there exists a spatial entity composed of just those two—however disparate those two may be." UN does allow, then, that there exist "gerrymandered composite entities"[7] such as the spatial entity composed of your cell phone and The Great Pyramid, the entity composed of The Great Pyramid and Lenin's nose, and the entity composed of Lenin's nose and Neptune.

Whereas Mereological Nihilism is counterintuitive because it implies that there are too few composite spatial entities—None!—UN is counterintuitive because it implies that there are too many. As Eric Olson would frame it, the concern is that, according to UN, "[t]he proportion of material objects that are dogs or bicycles or planets or anything else of interest" would be relatively small; and UN is thereby problematic given its implication that "[v]irtually all material things ... have completely arbitrary boundaries: they are merely 'ontological junk.'"[8] I presuppose that one should seek a theory of composition that is more restrained than UN—that restrains commitment to "ontological junk."

UN also implies counterintuitively that no spatial entity can come into being by way of composition (i.e., by way of the "assembling" of parts) and that no spatial entity can cease to be by way of *de*composition (i.e., by way of "losing" parts). For any spatial entity w composed of proper parts x and y, then w came into being at whatever time x and y first existed simultaneously: If x came into being *ex nihilo* on Monday somewhere on Earth, and if y came into being *ex nihilo* at noon Tuesday somewhere in the Andromeda Galaxy, then w would have come into being at noon Tuesday. But, given that, at noon Tuesday, x and y would have been 2.5 million light years away from one another, w's coming into being would not have been the result of the "assembly" or the "putting together" of x and y. Rather, w would have come into being at noon Tuesday because that is the time at which y came into being *ex nihilo* and thereby began its co-existence with x; and, per UN, the time at which x and y began co-existing would have been the time at which they began composing w.

By the same token, w cannot cease to exist by the "disassembly" or "taking apart" of x and y. If w is a materially solid spatial entity that is blown apart into separated parts x and y, and if, at noon Wednesday, x lands in a far corner of the Milky Way while y lands in Andromeda Galaxy, then, per UN, w would nonetheless exist at noon Wednesday even though x and y are 51.5 million light years away from one another. And, UN implies that w will continue to exist until x or y ceases to exist *in nihilum*.

A virtue of Conjoining, then, is that, unlike UN, Conjoining *does* allow both that there exist at least some nonoverlapping spatial entities that do *not* themselves compose a whole *and* that certain spatial entities cease to exist even if their (former) proper parts do not.[9]

Scattered objects. Presupposed is that one should accept a theory of composition that allows for the existence of "scattered" spatial entities—entities composed of "entirely separated" parts that are not "connected" (i.e., that do not "touch") directly or indirectly. Though one should resist UN's implication that *any* two spatial entities that are separated or scattered any which-a-way can compose a third, one should seek a theory of composition that leaves open the question of whether rocks and kidneys are composed of "scattered" atoms that are themselves composed of elementary particles that do not "touch"—particles that are "scattered in space."

Mereological Essentialism. Finally, presupposed without argument is the doctrine of Mereological Essentialism—the view that any spatial entity composed of parts is composed of its parts essentially:

ME For every x and y, if x is ever a proper part of y, then y is necessarily such that x is a proper part of y at any time that y exists.[10]

If a particular chair leg and proton are proper parts of your chair, then that particular chair has *that* leg and *that* proton as parts at every time that *that* chair exists. If, at a given time, you replace the leg or proton with a new leg or proton, then you may at that time continue to have a chair, but the object that is then your chair will not be *identical* with the object that *was* your chair.

In 1981, I formulated Conjoining in terms of the concept of no object's *lying between* two others—a concept that I accepted as primitive. I then claimed that two nonoverlapping spatial entities compose a whole when, and only when, they are *conjoined*—when and only when (i) no third entity lies between the two or when (ii) each has a proper part such that no third entity lies between those two proper parts. Now, by endorsing Spatial Directionalism, the concept of no spatial entity's lying between others need not be taken as a primitive concept; and unlike the 1981 formulation, the directionalist reformulation of Conjoining dodges the charge of circularity.

The concept of no spatial entity's lying between others is analyzable in terms of spatial directional relations:

D5.1 No spatial entity lies between x and y at time t =Df (i) x and y are nonoverlapping spatial entities; and (ii) for any spatial directional relation d that x bears to y at time t, there is no spatial entity z such that x bears d to z and z bears d to y.[11]

Two spatial entities are *directly conjoined* when there is no third spatial entity between them:

D5.2 x is directly conjoined with y at time t =Df No spatial entity lies between x and y at time t.

If a *materially solid* sphere is composed of two hemispheres, the hemispheres would be directly conjoined: There would be *no* spatial entity between them. If a hydrogen atom is a spatial entity composed of a single electron that spins around a nucleus, then the atom's electron and nucleus would not be "in contact" (as the materially solid sphere's hemispheres would be), but the electron and nucleus *would* be directly conjoined given that no third spatial entity would lie between them.

The examples involving the sphere and atom betray my ongoing commitment to the following principle of composition, which allows that the two hemispheres *do* compose a sphere and that the electron and nucleus *do* compose an atom:

> If x is directly conjoined with y, then there is a spatial entity composed of x and y.

Though being directly conjoined is a sufficient condition for composition, it is not a necessary condition: I continue to presuppose (as I did in 1981) that some spatial entities compose others even when they are *partially* (or *indirectly*) conjoined rather than *directly* conjoined. Suppose that a materially solid goblet's left half is composed of the left half of the goblet's cup and the left half of the goblet's stem, and similarly for the goblet's right half. I assume that the goblet-halves would compose a goblet even when the goblet is filled to the brim with materially solid liquid. When the goblet's cup is filled with liquid, the goblet-halves would not be *directly* conjoined given that the liquid would lie (partially) between the cup-halves. The goblet-halves, however, would nonetheless be *partially* conjoined: Each goblet-half would have a proper part that would be directly conjoined with a proper part of the other half. The stem's left half would be a proper part of the goblet's left half, the stem's right half would be a proper part of the goblet's right half, and the stem's left and right halves would be *directly* conjoined. Two spatial entities, then, are *partially conjoined* when they have proper parts that are *directly conjoined*:

> D5.3 x is partially conjoined with y at time t =Df At time t, x and y are non-overlapping spatial entities such that (i) it is false that no spatial entity lies between x and y, (ii) there exists a spatial entity w that is a proper part of x, (iii) there exists a spatial entity z that is a proper part of y, and (iv) w is directly conjoined with z.

The stem's left and right halves are directly conjoined but not, per D5.3, partially conjoined: Though each stem-half would have a proper part (i.e., a bottom-stem-half) that *would* be directly conjoined with a proper part of the other stem-half, it would be *true* (not false) that no spatial entity lies between the two stem-halves. Thus, the directly conjoined stem-halves fail to satisfy clause (i) of D5.3.

Two spatial entities are *conjoined* when they are either directly or partially conjoined:

> D5.4 x is conjoined with y =Df x is directly conjoined with y, or x is partially conjoined with y.

This revised analysis of the concept of conjoining allows a precise formulation of the Conjoining theory of composition. According to Conjoining, the answer to the Special Composition Question is that two spatial entities compose a whole when the two are *conjoined*:

> CON There exists a spatial entity composed of x and y, if and only if, x is conjoined with y.

Being conjoined, then, is both a necessary and sufficient condition for composition. Consider four examples that illustrate CON's implications.

First, per CON, a materially solid cube would be a spatial entity composed of its directly conjoined upper and lower halves: Those two cube-halves would be (per D5.2) directly conjoined given that no spatial entity lies between them. Second, if a hydrogen nucleus and electron are nonoverlapping spatial entities and if there is no spatial entity between the nucleus and electron, then (per D5.2) they would be directly conjoined, and (per CON) there would thereby exist a spatial entity (viz., a hydrogen atom) that would be composed of those two proper parts. Third, a materially solid hamburger would count as a spatial entity composed of the bun's top half and the spatial entity composed of both the burger and the bun's bottom half. With no spatial entity between them, the burger and bottom half of the bun would be (per D5.2) directly conjoined and would thereby compose (per CON) a spatial entity that would itself be directly conjoined with the top half of the bun. Assuming that the diameter of the burger is slightly larger than the diameter of the bun-halves, it would be false, however, that there would be a spatial entity that is a whole bun that would be a proper part of the sandwich: Because the burger would (entirely) lie between the top and bottom bun-halves, the two halves would be neither directly nor partially conjoined and thereby would *not* compose a "scattered" spatial entity—there would *not* be a whole bun that is then "scattered."

Finally, consider materially solid U-shaped ice tongs with an ice cube pinched between the ice tongs' tips. *Does* there exist a spatial entity (viz., the U-shaped ice tongs) that is composed of the two arms with tips that pinch the ice cube? Given that the ice cube lies between the ice tongs' two arms, the two arms would not be *directly* conjoined. The two arms, however, *would* be *partially* conjoined given that each arm has a proper part that would be directly conjoined with a proper part of the other: At the bend of the ice tongs—the ice tongs' fulcrum—there would be a proper part of one arm that would be in "direct contact with" a proper part of the right arm; and given that no spatial entity would lie between those two proper parts, they would thereby be directly conjoined.[12] If the ice cube lies "entirely between" the ice tongs' two tips, then the ice tongs' tips would be neither directly nor partially conjoined and would thereby *not* compose a "scattered object" composed of just the two tips.

IMPLICATIONS

Mereological simples. Conjoining allows for the possibility that *part*-less and dimensionless *mereological simples (atoms, simplons)* exist.[13] Mereological simples would have no proper parts; so, one can characterize mereological simples as those spatial entities that, per D3.7, can exhibit internally *no* spatial directional relations:

> D5.5 x is a mereological simple =Df x is a spatial entity that is necessarily such that there is no spatial directional relation that x exhibits internally at any time that x exists.

Although no spatial directional relation could obtain between a mereological simple's nonoverlapping proper parts (given that simples *lack* proper parts), spatial directional relations *could* obtain between a mereological simple and some other spatial entity. Thus, per D3.3, it *is* possible that mereological simples exhibit spatial directional relations *externally*.

Not only is CON consistent with the existence of mereological simples, but Conjoining allows for the possibility that there exists a spatial entity composed of nothing but two directly conjoined simples. CON also allows for the possibility that there exist both a three-dimensional materially solid, nonscattered spatial entity composed of indefinitely many simples *and* a three-dimensional nonsolid, scattered spatial entity composed of a swarm of a finite number of simples.

Three-dimensional materially solid wholes. A *materially solid* spatial entity would be an entity composed of parts that are *connected*—parts that are *entirely* or *partially connected* in such a way that nothing—"not even a gap"—lies between them. The relevant concepts can be explicated in terms of possible locations (i.e., in terms of certain sets of spatial directional relations).

Two spatial entities are *entirely connected* ("are completely in direct contact") when no possible location lies between *their* locations:

> D5.6 q is a possible location that lies between possible locations p and r =Df For any time t, p, q, and r are necessarily such that if p is the actual spatial location for something x, q is the actual location for something y, and r is the actual location for something z, then there is at least one spatial directional relation d such that x bears d to y and y bears d to z.

> D5.7 x and y are entirely connected at time t =Df At t, there is a spatial entity w that is composed of x and y, and there is no possible location that lies between the actual spatial locations for x and y.

Given that there would be no possible location between the actual spatial locations of a materially solid sphere's hemispheres, D5.7 implies that the hemispheres *would* be entirely connected. D5.7 also implies that the arms of materially solid ice tongs arms would *not* be entirely connected with one another: Between the actual spatial locations of the ice tongs' indirectly conjoined arms, there are many possible locations. Some of those possible locations serve as the actual locations of the ice cube and its proper parts, and others may serve as the actual locations of air molecules; but none serves as the actual spatial location of a proper part of the ice tongs. The arms of the ice tongs, then, would be *partially connected*:

> D5.8 x and y are partially connected at time t =Df At t, there is a spatial entity w such that (i) w is composed of x and y, (ii) there is a proper part of x and a proper part of y that are entirely connected, (iii) there is a possible location p that lies between the actual spatial locations for x and y, and (iv) there is no proper part of w for which p is the actual spatial location.

In virtue of being, respectively, entirely connected and partially connected, both the materially solid sphere's hemispheres and the materially solid ice tongs' arms would be *connected*:

> D5.9 x and y are connected at time t =Df At time t, x and y are entirely connected or partially connected.

Spatial entities that are *materially solid* are those entities composed of parts that are connected:

> D5.10 w is a materially solid spatial entity at time t =Df For any x and y at time t, if w is composed of x and y, then x and y are connected.

No stand is taken on whether it *is* logically possible that there exist materially solid spatial entities composed of indefinitely many mereological simples. CON does, however, leave open this possibility.

Scattered spatial entities. In defending the thesis that belief in scattered objects is "beyond reasonable doubt," Richard Cartwright writes:

> If natural scientists are to be taken at their word, all the familiar objects of everyday life are scattered.... There is at the moment a pipe on my desk. Its stem has been removed, but it remains a pipe for all that; otherwise no pipe could survive a thorough cleaning. So at the moment the pipe occupies a disconnected region of space.... Consider, finally, some printed inscription: the token of 'existence' on the title page of my copy of McTaggart's *The Nature of Existence*, for

example. Presumably it is a material object—a "mound of ink", as some say. But evidently it occupies a disconnected region of space.[14]

In his commentary on Cartwright's work, Chisholm is friendly toward the existence of scattered objects, noting that "the following would seem to be ontologically respectable: the United States, the solar system, a suite of furniture, a pile of coal, a watch that is spread out on the watch repairer's workbench, printed words, and lowercase letters i and j, the constellation Cassiopeia."[15] Following Cartwright and Chisholm, I characterize scattered spatial entities as those spatial entities composed of parts that are not connected:

D5.11 w is a scattered spatial entity at time t =Df At t, there exist two spatial entities x and y such that w is composed of x and y, but x and y are not connected.

If there exists a hydrogen atom composed of an electron orbiting around the nucleus, D5.11 would imply that the hydrogen atom is a scattered spatial entity: The electron and nucleus would not be connected—they would be neither entirely nor partially connected. Whether mereological simples exist is left open, but D5.11 does allow for the possibility that there exists a scattered spatial entity composed of two mereological simples or of one mereological simple and one materially solid spatial entity.

D5.11 does *not* allow a materially solid doughnut (in a vacuum) to count as a scattered spatial entity even though there is a "completely empty gap" or "completely empty hole" between proper parts of the doughnut: The left C-shaped half-doughnut *is* partially connected with the right half-doughnut. The doughnut, however, *would* have a proper part that is a scattered object: The quarter-doughnut to the left of "the hole" and the quarter-doughnut to the right of "the hole" would not be connected; but in virtue of their being directly conjoined, they would compose a spatial entity that *would* be a proper part of the materially solid doughnut.

Having in hand a reformulation of Conjoining *and* analyses of concepts of various spatial entities that may exist, I devote the remainder of this chapter to the formulation and evaluation of objections to Conjoining, beginning with van Inwagen's "circularity" objection.

THE "CIRCULARITY" OBJECTION

van Inwagen has objected that certain proposed answers to the Special Composition Question are circular. After a sketch of what "special bonding" answers to SCQ involve and after an explanation of why "special bonding" theories of composition *and* the 1981 version of Conjoining are all subject to van Inwagen's charge of circularity, this section closes with an explanation of why CON (the reformulation of Conjoining) is *not* circular.

van Inwagen observes that one may be tempted to answer SCQ with an appeal to "multigrade relations of physical bonding," claiming that entities compose a whole when the entities bear the right sort of "bonding" relations to one another.[16] A formulation of a "special bonding" theory of composition would have something like the following structure, where 'P-bonding' and 'A-bonding' refer to the types of physical bonding that can occur, respectively, among *part*-less particles and among atoms (e.g., carbon atoms):

> SB Discrete entities compose something *y*, if and only if, (i) the entities are particles not composed of parts and are maximally P-bonded, or (ii) the entities are atoms composed of parts and are maximally A-bonded.

SB implies that certain *part*-less particles would compose a whole when those particles are maximally bonded in the way that mereologically simple particles can be bonded. And SB also implies that carbon, oxygen, and hydrogen atoms—atoms that *have* parts—would compose a whole when those atoms are maximally bonded in the way that such *non*-mereologically simple atoms can be bonded. Presumably, one would turn to chemists or physicists for more illuminating accounts of what such bonding would involve.

Rightly so, van Inwagen objected that SB-type theories of composition are circular: The conditions for when it is that entities *do* compose a whole are cast in terms of physical bonding—(i) the type of physical bonding that obtains among things that are *composed* of entities *not* themselves *composed* of parts (e.g., mereological simples), and (ii) the type of physical bonding that obtains among things that are *composed* of entities (e.g., carbon atoms) that *are* themselves composed of parts. In short, charges van Inwagen, in formulating the necessary and sufficient "bonding" conditions for composition, the SB theorist does presuppose conditions for composition when identifying the type of bonding exhibited by spatial entities *not* composed of parts and when identifying the second type of bonding exhibited by spatial entities that *are* composed of parts. To specify the spatial entities that can exhibit P-bonding, the SB theorist presumes that simples *are* things *not* composed of parts; and to specify the spatial entities that can exhibit A-bonding, the SB theorist presumes that atoms (e.g., carbon atoms) are things that *are* composed of parts.

To appreciate the force of van Inwagen's "circularity" objection and its implication for the 1981 version of Conjoining, consider a "Special Moral

Question": "When is it that a given action *is* morally obligatory?" To skirt the standard objection that maximizing utility would justify the breaking of a promise (when promise-breaking would produce just a bit more utility than keeping one's promise), a naïve utilitarian may be tempted by the following answer to SMQ:

> UT   An action A is morally obligatory, if and only if, (i) no other action would produce a greater utility than A would produce, and (ii) performing A is not immoral.

On behalf of UT, the naïve utilitarian could note that, when promise-breaking produces a bit more utility than promise-keeping, UT would *not* imply that promise-breaking is morally obligatory given that promise-breaking *is* immoral and thereby would not satisfy the second clause of UT. UT, however, is a blatantly circular answer to SMQ: To specify the type of action that *would* be morally obligatory if it maximizes utility, this naïve utilitarian presupposes an unspecified condition for obligatoriness when stipulating that a utility-maximizing action would be obligatory *only if* its nonperformance is not itself *morally obligatory*. This naïve utilitarian formulates the necessary and sufficient conditions for moral obligatoriness in terms of moral obligatoriness.

Just as UT is a circular answer to the Special Moral Question, van Inwagen would charge the defender of SB with circularity: By presupposing that certain things subject to a particular kind of bonding lack composing parts while certain other things subject to a different kind of bonding *have* composing parts, the defender of SB is guilty of formulating the necessary and sufficient conditions for composition in terms of composition. van Inwagen's "circularity" objection applies not only to "special bonding" answers to SCQ but to the version of Conjoining implicit in my 1981 work. After formulating this original version, I will explain why it, like SB, is subject to van Inwagen's "circularity" objection.

In my original formulation of Conjoining, I used two of Chisholm's definitions, taking the concept of a proper part as primitive:[17]

> D_a x is discrete from y [x and y are nonoverlapping] =Df (i) x is other than y, (ii) there is no z such that z is a proper part of y and z is a proper part of x, and (iii) x is not a proper part of y and y is not a proper part of x.

> D_b w is strictly made up of x and y =Df (i) x is a proper part of w, (ii) y is a proper part of w, (iii) x is discrete from y, and (iv) no proper part of w is discrete both from x and from y.

I then analyzed the concept of direct conjoining in terms of the concept of lying between, which (as noted earlier) was also taken as primitive:

D_c x is directly conjoined with y =Df x is discrete from y and there is no third object which lies between x and y.

D_d w is conjoined with z =Df There is an object x and an object y such that (i) x is part of w and y is part of z and (ii) x is directly conjoined with y.

D_e w is strictly made up of x and y in virtue of x and y being conjoined =Df (i) w is strictly made up of x and y and (ii) and either x is directly conjoined with y or x is conjoined with y.

In 1981, I did not explicitly formulate a theory of composition; instead, I offered D_e as an analysis of the concept of a whole's being made up of (i.e., being composed of) *conjoined* parts. Implicit, then, was the following formulation of Conjoining:

CON* Nonoverlapping [discrete] spatial entities compose a spatial entity, if and only if, those nonoverlapping spatial entities are conjoined [per D_d].

Consider why, exactly, CON* *is* subject to van Inwagen's "circularity" objection.

CON* is an attempt to specify the necessary and sufficient conditions for when it is that a spatial entity *is* composed of other spatial entities. And, as van Inwagen would put the point, "the right-hand constituent of an answer to the Special Composition Question may not contain 'part' or 'compose' or any other mereological term."[18] That is, one should not (circularly) formulate the necessary and sufficient conditions for composition in terms that themselves presuppose necessary and sufficient conditions for composition. CON* violates this caveat: The conditions for composition are formulated in terms of directly conjoined and conjoined nonoverlapping spatial entities (per D_e)—spatial entities not *composed* of parts (e.g., mereological simples) or *composed* spatial entities that have no *composing* parts in common. To put the point epistemically, to know whether two nonoverlapping spatial entities compose a whole, one would have to know whether they are nonoverlapping spatial entities that are directly conjoined or conjoined; but to know whether a spatial entity does or does not overlap with another, one would have to know (i) whether the spatial entities are part-less—whether there exist *no* spatial entities that satisfy principles of composition to *compose* them—or (ii) whether the spatial entities *are* composed spatial entities such that no spatial entity that *composes* one also *composes* the other.

Why the revised formulation of Conjoining *is* not *subject to the "circularity" objection.* CON is not circular: Unlike CON*, the *explanans* of CON is *not* formulated in terms of mereological concepts that presuppose unspecified conditions for composition. Per D5.4, the concept of conjoining *is* analyzed in terms of *partial* and *direct* conjoining; and these concepts (D5.2 and D5.3) are analyzed in terms of the concept of no spatial entity's lying between others and the mereological concepts of a proper part and of nonoverlapping spatial entities. Ultimately, however, these mereological concepts are analyzed in terms of the concept of a spatial directional relation, which is *not* a primitive mereological concept! The concept of no spatial entity's lying between others is analyzed [per D5.1] in terms of nonoverlapping spatial entities, and the concept of nonoverlapping spatial entities is analyzed [per D3.6] in terms of a spatial entity's proper parts. By embracing spatial directional relations, the Spatial Directionalist can insist that the concept of a proper part need *not* be construed as a primitive mereological concept, but can instead be analyzed [per D3.5] in terms of spatial entities and their spatial locations; and the concepts of spatial entities and of their locations can themselves be analyzed [per D2.1 and D3.2] in terms of spatial directional relations and sets of such relations.[19]

Summary. To ask the Special Composition Question is to ask for a theory of composition that explains whether, at a given time, two spatial entities do or do not then compose a spatial entity of which the two would be parts. If a given theory of composition can preserve commonsense intuitions regarding which spatial entities do and do not compose others and which spatial entities are and are not themselves composed of parts, then that theory would have a mark in its favor. And, if the theory can also preserve scientists' reports regarding the composition of macro-objects by molecules, the composition of molecules by atoms, and the composition of atoms by yet smaller particles, then the theory would have another mark in its favor. With respect to these two standards, Conjoining ranks high: While skirting the "circularity" objection, Conjoining preserves the commonsense view that there exist kidneys, boulders, and planets, and that these are all (scattered) spatial entities composed of scattered molecules, of scattered atoms, and (ultimately) materially solid particles or particles that are mereologically simple. Another mark in favor of Conjoining is that it preserves the commonsense view that there are spatial entities that come into being and pass away by the assembling and disassembling of parts: One brings a table into being by attaching legs to a tabletop, bringing it about that no spatial entity lies between each leg and the tabletop; and, a demolition team destroys a building by bringing it about that many parts of the building cease to be conjoined: The various bricks, boards, and studs that now compose a pile of rubble have ceased to compose walls, floors, and ceilings.

Unfortunately, there may be *no* answer to SCQ that is entirely satisfying: *Every* theory of composition may have at least *some* counterintuitive implications. In the end, however, *some* counterintuitive implications are a small price to pay for a theory that preserves the commonsense views that there *do* exist spatial entities composed of conjoined parts, that spatial entities can come into being (when spatial entities are conjoined to compose a whole), and that spatial entities can cease to be (when their parts cease to be conjoined). At the very least, CON is far less counterintuitive than a view that implies that there exist no (composed) kidneys, boulders, or planets or that there *does* exist a spatial entity composed of a snake in Denver and the top half of Jupiter's largest moon.

Consider now a series of objections to Conjoining that trade on exposing counterintuitive implications of the theory. In each case, the defender of CON has an available explanation that renders the implication more palatable.

THE "ODD OBJECTS" OBJECTION

Though van Inwagen raised an "odd objects" objection with respect to an answer to SCQ other than Conjoining, consider an interpretation of his objection that targets CON:[20]

> Suppose that we shake hands. Does a new thing come into being at the moment our hands are conjoined? At that moment when no third object lies between us, does there come into being a thing shaped like a statue of two people shaking hands, a thing that has you and me as proper parts and that will perish when we cease to be conjoined by the handshake? Is there a spatial entity that fits just exactly into the region of space that you and I jointly occupy? No. But, surely, if you and I composed something when we were shaking hands, it would have all of these features. Therefore, despite our being conjoined, there would be nothing that you and I compose when we shake hands.[21]

The "odd objects" objection is this:

OO 1. If two humans, H1 and H2, shake hands at time t, then, at t, there would be no third object between H1's right hand and H2's right hand.
    2. H1's right hand is a proper part of H1, and H2's right hand is a proper part of H2.
    3. If (i) Conjoining is correct, (ii) at t there is no third object between H1's right hand and H2's right hand, and (iii) H1's right hand is a proper part of H1, and H2's right hand is a proper part of H2, then H1 and H2 would be conjoined at time t.[22]
    4. If H1 and H2 are conjoined at time t, then an object O composed of H1 and H2 would come into being at time t, O would be shaped like a statue of two people shaking hands, and O would cease to exist when H1 and H2 cease to be conjoined.

5. If Conjoining is correct and if H1 and H2 shake hands at time t, then, at time t, an object O composed of H1 and H2 would come into being, O would be shaped like a statue of two people shaking hands, and O would cease to exist when H1 and H2 cease to be conjoined. (from 1,2,3,4)
6. It is false that H1 and H2 shake hands at time t only if there would come into being at t an object O shaped like a statue of two people shaking hands such that O would cease to exist when H1 and H2 quit shaking hands.
7. Therefore, Conjoining is not correct. (from 5,6)

The "odd objects" objection turns on OO-6, but why suppose that this premise is true? Is it because the handshaking entity would be short-lived, coming into being and passing away in a matter of seconds? Surely not: Other spatial entities composed of parts are short-lived but are nonetheless wholes for as long as they exist. For example, a child may construct a 16-block tower by stacking alphabet blocks one atop another. The stable 15-block tower may exist but for a second or two before the child knocks it apart, scattering blocks across the floor.

One should not endorse OO-6 on grounds that we do not have (at least in English) a simple term like 'human,' 'hand,' or 'statue' that we use to refer to short-lived spatial entities that are statue-like and composed of handshaking humans. What spatial entities exist—a metaphysical question—is not a function of some arbitrary linguistic fact about what fine-grained nouns a particular language happens to contain. Moreover, although the English language contains no term for such an entity, the defender of OO would presumably countenance the existence of a massive enduring spatial entity composed of a metal cube, pyramid, and sphere blown together by a tornado and fused by a random lightning bolt. And, the fused cube, pyramid, and sphere would compose the massive spatial entity even if a second lightning bolt blows it apart a few seconds after the entity came into being.

Conclusion. It is reasonable to believe that the universe includes innumerable odd objects that may be short-lived and that may come into being and pass away as various spatial entities become conjoined or cease to be conjoined by way of chance social interactions, by deliberate human activity, or by the random behavior of storms, ocean currents, or earthquakes. A virtue of Conjoining, then, is that it *does* allow, as an adequate theory of composition should, for the existence of "odd objects."

THE "IDENTITY" OBJECTION

Consider a version of van Inwagen's "identity" objection, modified to target Conjoining:[23]

Suppose that I touch your shoulder with my elbow. Would the spatial entity that comes into being with the conjoining of your shoulder and my elbow be the same entity that came into being when we earlier shook hands? Could Conjoining allow a non-arbitrary answer to this question?

Below is a more formal version of the "identity" objection, formulated with the assumption that H1 and H2 undergo no mereological change between times t and t':

ID 1. Two humans, H1 and H2, shake hands at time t only.
2. H1 touches H2's shoulder with H1's elbow at another time t' only.
3. If Conjoining is correct and if two humans, H1 and H2, shake hands at time t, then spatial entity SE1 composed of H1 and H2 would come into being at time t.
4. If Conjoining is correct and if H1 touches H2's shoulder with H1's elbow at time t', then spatial entity SE2 composed of H1 and H2 would come into being at time t'.
5. Therefore, if Conjoining is correct, then SE1 composed of H1 and H2 comes into being at time t and SE2 composed of H1 and H2 comes into being at time t'. (from 1,2,3,4)
6. If SE1 composed of H1 and H2 comes into being at time t and if SE2 composed of H1 and H2 comes into being at time t', then there is some nonarbitrary explanation for why SE1 = SE2 or why SE1 ≠ SE2.
7. Therefore, if Conjoining is correct, then there is some nonarbitrary explanation for why SE1 = SE2 or why SE1 ≠ SE2. (from 5,6)
8. There is *no* nonarbitrary explanation for why SE1 = SE2 or why SE1 ≠ SE2.
9. Therefore, Conjoining is not correct. (from 7,8)

The "identity" objection turns on ID-8, and there is good reason to reject it. Although it *is* true that Conjoining does not alone explain whether SE1 = SE2 or SE1 ≠ SE2, coupling CON with Mereological Essentialism (ME) *does* allow a nonarbitrary reply to the "identity" objection.

At t, SE1's proper parts would include H1, H2, H1's extended hand, H2's extended hand, *and* (per CON) the spatial entity composed of the two directly conjoined handshaking hands. ME implies that SE1 at t would *not* be identical with SE2 at t' if SE1 has proper parts that SE2 lacks or vice versa. And indeed, SE1 would have at least one proper part that SE2 lacks, and SE2 would have a proper part that SE1 lacks. The two directly conjoined handshaking hands would compose a spatial entity D that would be part of SE1 at t, but composed entity D would not be a part of SE2 at t' because D would have ceased to exist when the two hands separated and ceased to be directly conjoined when air molecules came to lie between the hands. Similarly, at t', SE2 would have a proper part E composed of H1's elbow and H2's shoulder, but E was not a proper part of SE1 at t' because E did not exist at t': H1's elbow and H2's shoulder were not conjoined at t' and thereby did not then compose E.

Conclusion. If spatial entities *are* composed of parts that are conjoined (or that "touch" or that are "in contact"), then an adequate theory of composition should allow for a nonarbitrary explanation of whether a composed spatial entity at one time is or is not identical with a composed spatial entity at some other time. A virtue of Conjoining is that, when coupled with ME, it does allow for nonarbitrary answers to such identity questions.

THE "TWO BEGINNINGS" OBJECTION

Below is a version of van Inwagen's "two beginnings" objection, modified to pose a problem for CON:[24]

> If by shaking hands, our hands are directly conjoined and thereby compose a spatial entity, does that same spatial entity come into (or resume) existence every time that we shake hands? One would like to believe that this question has a nonarbitrary answer.

Below is a more formal version of the "two beginnings" objection, formulated with the assumption that neither H1 nor H2 undergoes any mereological change between times t and t':

TB 1. Two humans, H1 and H2, shake hands at time t and again at t'.
    2. If Conjoining is correct and if H1 and H2 shake hands at time t and again at t', then a spatial entity SE1 composed of H1 and H2 comes into being at t and a spatial entity SE1' composed of H1 and H2 comes into being at t'.
    3. Therefore, if Conjoining is correct, then a spatial entity SE1 composed of H1 and H2 comes into being at time t and a spatial entity SE1' composed of H1 and H2 comes into being at time t'. (from 1,2)
    4. If a spatial entity SE1 composed of H1 and H2 comes into being at time t and if a spatial entity SE1' composed of H1 and H2 comes into being at time t', then there is some nonarbitrary explanation for why SE1 = SE1' or SE1 ≠ SE1'.
    5. Therefore, if Conjoining is correct, then there is some nonarbitrary explanation for why SE1 = SE1' or SE1 ≠ SE1'. (from 3,4)
    6. There is *no* nonarbitrary explanation for why SE1 = SE1' or SE1 ≠ SE1'.
    7. Therefore, Conjoining is not correct. (from 7,8)

The "two beginnings" objection turns on TB-6—on whether there is a nonarbitrary explanation for whether a spatial entity can have two beginnings—whether a spatial entity can come into being and then cease to be and *then* come into being a second time. Neither Conjoining nor Conjoining coupled with ME can, alone, provide an answer to this question, but this does not imply that TB-6 is true after all. Whether TB-6 is true turns on the independent metaphysical question of whether an object can begin to exist at a time after it ceases to be. What is obvious is that no object can come into being after the time at which that object ceases to exist *for the final time*. The question at hand, however, is whether it is necessarily true that the time at which a particular object ceases to exist *is* the final time at which *that* object ceases to exist. I take no stand on how *this* metaphysical question should be answered: CON is consistent both with its being possible that a composed spatial entity comes into being more than once and with its being impossible that a composed entity comes into being more than once.

THE "ARBITRARY BOUNDARY" OBJECTION

Eric Olson observes that Conjoining faces "all the disadvantages of Universalism and then some."[25] What, exactly, *are* these disadvantages, and are they compelling reasons for rejecting Conjoining? If there are disadvantages to be suffered, does Conjoining suffer fewer disadvantages than Universalism or some other theory of composition?

Consider Olson's complaint that Universalism allows for much "ontological junk"—objects with "completely arbitrary boundaries" that, unlike dogs or bicycles, are of no interest to us.[26] Indeed, like Universalism, Conjoining also allows for the existence of spatial entities with "arbitrary boundaries"— entities that are of no interest to us. For example, Universalism allows for the existence of a scattered object composed of the southernmost proton that is a part of Key West and the northernmost neutron that is a part of Point Barrow. Though Conjoining disallows the existence of *that* "ontological junk" (because, with many spatial entities lying between them, the proton and neutron would not be "conjoined"), Conjoining *does* allow for the existence of a scattered entity composed of a space-walking astronaut and the Space Shuttle's forward-most rivet if, in outer space, no spatial entity lies between those two things. Presumably, Olson would describe the scattered astronaut/rivet as "ontological junk."

One who defends CON should note that whether one *conceives* a given spatial entity as "ontological junk" is irrelevant to whether it *is* a spatial entity composed of parts: Regardless of whether the astronaut/rivet's boundary is or isn't "arbitrary" and regardless of whether the astronaut/rivet would be of any interest to anyone, the relation that obtains between the astronaut and the rivet is the same relation that obtains between a hydrogen atom's nucleus and its electron—a nonarbitrary relation that captures a plausible sense in which those two entities are *conjoined* and thereby compose a scattered entity.

Olson may insist that this reply does not take seriously enough the troublesomeness of "arbitrary boundaries" vis-à-vis "ontological junk." For example, what *is* the boundary of the astronaut/rivet? Would it be the sum of a certain "outer" part of the astronaut and a certain "outer" part of the rivet? But which parts, exactly, of the astronaut and rivet would constitute the sum that is the scattered entity's boundary? Some may complain that such a sum would not "surround" the scattered entity—it would not "enclose" the two composing entities *and* the "gap" between them. Others might object that this "sum" view would not allow that the astronaut/rivet's outer boundary grows larger as the astronaut and rivet move further apart. If, however, the astronaut/rivet's outer boundary is not a sum of certain "outer" parts of the astronaut and of the rivet, then what could the boundary possibly *be*? Could it be a particular region of *space*? (But *which* region? And, wouldn't this involve ontological commitment to substantivalist space after all?) Would the boundary be a metaphysically bizarre dependent particular that "surrounds" the astronaut and rivet as if someone had drawn a line enclosing these two spatial entities and the "gap" between them? But of *what* would such a line-like boundary be composed? And how could it expand and contract as the astronaut moves away from, and then closer to, the rivet? Would the line-like boundary "fit snugly around" the astronaut and rivet like a snug belt around one's waist? Would the line-like boundary instead be like a belt that is a bit too big, passing around the astronaut and rivet around with a small margin between the boundary and the two spatial entities? (How wide would this margin be?) Or, would the line-like boundary be ellipse-like or hourglass-shaped—either bulging outward away from "the space" between the astronaut and rivet or dipping inward directly between these two entities?

Below is a formal version of the "arbitrary boundary" objection:

AB 1. If Conjoining is correct, then there exists "ontological junk"—there exist spatial entities with boundaries of unknown composition that lack nonarbitrary shape and size with respect to how they "bound" the spatial entities of which they are the boundaries.
2. It is dubious that there exist spatial entities with boundaries of unknown composition that lack nonarbitrary shape and size with respect to how they "bound" the spatial entities of which they are the boundaries.
3. Therefore, Conjoining is dubious.

First, as noted earlier, the "arbitrary boundary" objection poses a problem not only for CON but for Universalism and for any other theory of composition that allows for the existence of *scattered* spatial entities that some would count as "ontological junk." Though a view that allows for "ontological junk" may be dubious, an alternative may be yet more costly if it eliminates "ontological junk" at the expense of eliminating commonplace scattered spatial entities as well—spatial entities such as hydrogen atoms, water molecules, kidneys, rocks, and planets.

Second, AB-2 is not itself objectionable: One *should* be wary of countenancing mysterious boundaries of unknown composition and of arbitrary shape and size. But there is hope that the defender of CON can endorse AB-2 and block the "arbitrary boundary" objection by *rejecting* AB-1 with an appeal to a broad, nonmysterious theory of boundaries for both scattered and nonscattered spatial entities. I formulate such a theory of boundaries in the following chapter, suggesting that the boundaries of scattered and nonscattered spatial entities are certain nonarbitrary (directionalist/relationalist) spatial locations. Thus, without ontological commitment to substantivalist space or to mysterious dependent particulars that "bound" scattered and nonscattered spatial entities, coupling CON with Spatial Directionalism may allow both that there exist hydrogen atoms and other scattered spatial entities *and* that scattered entities have nonarbitrary boundaries.

THE "BRICK-AND-MORTAR" OBJECTION

Olson mentions in passing yet another objection to Conjoining:

> Hestevold suggests . . . that things compose something just when there is no material thing between them. . . . So the atoms of my house compose something, but the bricks of my house don't, because they are separated by mortar.[27]

The "brick-and-mortar" objection that Olson suggests is this:

BM 1. The atoms of a house compose something (viz., the house) only if its bricks also compose something.
    2. Therefore, it is not true both that the atoms of a house compose something, and the bricks do not. (from 1)
    3. If Conjoining is correct, then the atoms of a house compose something but the (mortar-separated) bricks do not.
    4. Therefore, Conjoining is not correct. (from 2,3)

BM-3 is true: Olson is right that, per Conjoining, a house's atoms would compose something (viz., the scattered entity that would be the house); and it is false that the bricks themselves would compose a (scattered *or* nonscattered) spatial entity given that there is mortar that lies between any two bricks. The "brick-and-mortar" objection turns, then, on its first premise. Though Olson apparently believes that BM-1 is obvious, consider a reason to reject it.

First, it *is* reasonable to believe that a house's atoms *would* compose something—viz., the house—and CON can explain why this is so. The house's parts would include not only (scattered) atoms but parts that are atomic nuclei, parts that are molecules, parts that are bricks, and parts that are chunks of mortar. Because these various spatial entities *would* be conjoined in various ways, they would (per CON) thereby compose a scattered entity that would be the house. For example, a nucleus and several electrons may compose a calcium atom while other nuclei and electrons compose silicon atoms, and various calcium, silicon, and oxygen atoms may compose various silicate molecules that themselves compose bricks and mortar. Although it is plausible to claim that the house's atoms compose something, it would be a mistake to claim that the house's atoms compose something only if the house's bricks compose something.

The reason that affirming BM-1 would be a mistake is that there is a relation that would obtain among any two of the house-parts that would *not* obtain between any two of the house's bricks. The house's parts would be assembled in such a way that, between any two of the house's nonoverlapping parts, there would exist nothing or nothing that is not itself a part of the house. For example, with respect to any two calcium nuclei that are parts of the house, those two nuclei would be directly conjoined (with nothing between them), or other parts of the house (e.g., other nuclei, bricks, or chunks of mortar) would lie between those two nuclei. With respect to the sum of the house's *bricks* alone, this relation would *not* obtain between any two bricks: Between any two bricks, there *would* exist something—a chunk of mortar!—and that something would *not* be a "part" of the sum of the bricks alone. In short, then, one can reject BM-1 by claiming plausibly that the house's parts satisfy composition conditions and thereby compose a whole whereas the scattered bricks alone fail to satisfy those composition conditions and thereby fail to compose a whole. Only if one were at least somewhat sympathetic with the Universalist's unbridled theory of composition would one take seriously the view that two bricks compose a spatial entity even when those two bricks are entirely separated by mortar.

Summary. The "brick-and-mortar" objection is not compelling because (i) there is no compelling reason to accept the counterintuitive implication of Universalism that *any* two spatial entities compose a whole, and (ii) there appears to be no plausible alternative to CON that allows for the composition of spatial entities by those that are *not* conjoined. Moreover, by endorsing CON and rejecting BM-1, one can preserve fundamental intuitions about when it is that particles, bricks, and mortar do and do not compose larger spatial entities.

IS THERE A MORE PLAUSIBLE MODERATE ANSWER TO SCQ?

One may find Conjoining more appealing if one resists the extreme answers to the Special Composition Question—if one resists both the Mereological Nihilist's view that there exist no composite spatial entities *and* the Universalist's view that any two spatial entities compose a third. Before endorsing CON, however, one should consider whether there is a nonextreme answer to SCQ that is more plausible than CON. Consider three possibilities.

Contact. van Inwagen has entertained and rejected *Contact*—the view that two entities compose something, when and only when, they are *in contact* with one another.[28] A virtue of Contact is that it does preserve the commonsense view that a sphere would come into being when the circular faces of two materially solid hemispheres come together and make contact. And Contact preserves the commonsense view that the sphere would cease to exist when the two hemispheres separate, ceasing to be in contact. Also, in its favor is that Contact preserves the view that there exists no material entity composed of The Great Pyramid and Lenin's nose given that those two material entities are not in contact.

Invoking the "Odd Objects" Objection, some could object that Contact is too loose. For example, van Inwagen observes that Contact implies counterintuitively that when two people quickly shake hands, there would temporarily exist a spatial entity composed of just those two handshaking people. This is not a compelling objection to Contact for the same reason that the "odd objects" objection is not a compelling objection to Conjoining: There is no good reason to deny that fifteen briefly connected building blocks compose a briefly existing tower or that two handshaking humans compose a briefly existing spatial entity.

Instead of objecting that Contact is too loose, I object that it is too restrictive: Contact implies that *no* scattered objects could exist—that there could exist no atoms composed of scattered subatomic particles, no molecules composed of scattered atoms, and no diamonds composed of carbon molecules. Though an adequate theory of composition should imply that The Great Pyramid and Lenin's nose do *not* compose a scattered spatial entity, an adequate theory *should* allow that at least some scattered spatial entities exist.

Another reason to resist Contact is that the concept of contact is itself vague: What, *exactly*, does the contact of two *nonoverlapping* spatial entities involve?[29] Contact would not involve the sharing of a common part: If two materially solid hemispheres share a common part (however thin that part might be) then those two hemispheres would *not* be *non*overlapping hemispheres. If a materially solid sphere has a proper part that is a complete materially solid hemisphere, then how could its other composing part also be a complete materially solid hemisphere that has a two-dimensional circular face that is *in contact with* the other hemisphere's circular face? Is it impossible to push two perfectly *complete* materially solid hemispheres together until they are *in contact* and thereby compose a perfect materially solid sphere? Is it impossible to halve a materially solid sphere such that its two left and right hemispheres would exist, but would cease to compose a sphere? (In chapter 6, this problem of contact arises again with respect to boundaries.)

Fastening. Philosophers have also entertained and rejected *Fastening*: "Two entities compose something, when and only when, they are fastened (bound) together in such a way that relatively few forces could cause the two to separate (i.e., to cease to be in contact) without damaging one or both of the composing entities."[30] Using van Inwagen's examples, this view does preserve the commonsense view that a nut-threaded-on-a-bolt and a wristwatch are both spatial entities composed of suitably bound parts whereas a child's tower of stacked ABC blocks is not: The nut-and-bolt and watch may well survive intact if subjected to a variety of different forces of various magnitudes and directions; but, applied to the stacked blocks, any one of these forces may be sufficient to cause the disintegration of the tower.

van Inwagen resists Fastening, charging that it is subject to the "odd objects" objection and is thereby too loose. van Inwagen complains that Fastening implies that there would exist a spatial entity composed of two humans when, while shaking hands, their hands become paralyzed such that neither can release the other's hand. Claiming that there would exist no such object composed of two handshaking humans, van Inwagen concludes that Fastening is mistaken. Again, for reasons explained earlier in this chapter, the "odd objects" objection is not compelling: Regardless of whether their hands are or are not bound in a death grip, there is no good reason to reject the view that two handshaking humans compose a spatial entity.

Instead of objecting that Fastening is too loose, I object that it is too restrictive: Some spatial entities *do* compose others even though their parts are *not* fastened together. For example, fifteen stacked ABC blocks *do* compose a tower even if the blocks are not sufficiently fastened together; and, though they are *not* fastened together, fifty-two stacked playing cards do compose a deck.

A second reason to resist Fastening is that the concept of fastening is itself vague. For the sake of developing this objection simply, ignore the fact that the effect of an applied force on a spatial entity is a function of both the force's magnitude *and* the direction from which the force is applied; consider magnitude only. Imagine a particularly fragile hand-made book constructed of a hundred delicately thin pages bound by a light application of paste along the pages' edges. Such a book *is* a spatial entity composed of parts, but the book could survive the application of relatively few forces given that the thumbing of a page could easily tear the page itself or could easily tear the page from the book. Thus, there would seem to be no specific minimum magnitude of force such that two spatial entities compose a third, if and only if, the third and its parts can survive intact the application of a force of that specific magnitude. This concern is magnified when one considers that, conceivably, the tower of blocks could survive the application of many relatively weak forces that would be sufficient for destroying the book. The minimum force required to tear the top page from the book may be less than the minimum force required to move one of the stacked ABC blocks. Is composition, then, a function of what *kinds* of entities are fastened? Could it be that there exists no spatial entity composed of lightly pasted thin pages whereas there exist many mass-produced books composed of stitched and glued heavy-weight pages? Do there exist towers composed of stacked blocks of granite but no towers composed of a child's ABC blocks? Do there exist houses composed of bricks that were fastened together with cement but no objects composed of two handshaking humans whose hands were fastened together with syrup?

Or, suppose that the stacked but easily disassembled ABC blocks are transported from Earth to Jupiter where the gravity is stronger. On Jupiter, the stacked blocks may survive forces sufficient for knocking them apart on Earth. Would the stacked blocks compose a tower on Jupiter but not on Earth? Does Fastening allow that composition is arbitrary, depending on *where* in the universe spatial entities happen to be located?

Finally, Fastening is problematic because it does not obviously allow for scattered spatial entities. *Does* the concept of fastening allow for the possibility that there exist a certain proton and neutron "appropriately fastened" to compose a spatial entity that is a one-proton one-neutron atomic *nucleus*? *Does* the concept of fastening allow for the possibility that there exists a scattered spatial entity that is a hydrogen *atom* composed of a one-proton one-neutron nucleus that is "appropriately fastened" with a certain electron? (And, in virtue of what conditions would *that* atom's nucleus be "appropriately fastened" to *that* electron rather than to some other neighboring electron?) *Does* the concept of fastening allow that water molecules *are* scattered spatial entities composed of "appropriately fastened" hydrogen and oxygen atoms that are themselves scattered spatial entities?

If Fastening does imply that a hydrogen atom is a scattered spatial entity composed of a one-proton one-neutron nucleus and an electron, then under what conditions would that nucleus and electron cease to be fastened and thereby cease to compose a spatial entity? If a force applied to the hydrogen atom moves the nucleus and electron closer together, would the force thereby constitute damage or deformation of the atom such that the atom would cease to exist—such that there would cease to exist a scattered spatial entity composed of a fastened nucleus and electron? Or, to put the problem another way, suppose instead that there exist a one-proton one-neutron nucleus on Neptune *and* an electron on Mercury; and suppose that this nucleus and this electron do not now compose a scattered spatial entity. What conditions must the nucleus and electron come to satisfy such that these two spatial entities *would* compose a scattered spatial entity that is, say, a hydrogen atom? Must these two scattered spatial entities move closer together before they can compose a scattered spatial entity? If so, at what distance would the hydrogen atom come into being by way of fastening? (And why *this* distance rather than some other distance a tiny bit greater or a tiny bit smaller?) If there is one specific distance relevant to the fastening of the nuclei and electrons of scattered hydrogen atoms, are there yet other distances that are relevant to the fastening of the nuclei and electrons of other types of scattered atoms? If so, why would the nucleus and electrons of, say, a helium atom become fastened at a different distance than the nucleus and electrons of, say, a carbon atom? If there is but a single specific distance at which any nucleus becomes fastened to an electron, then in virtue of what is fastening a function of that specific distance rather than some other specific distance? Without answers to such questions and a crisp account of what fastening involves, Fastening is *not* a plausible answer to the Special Composition Question.[31]

Living Organism. After entertaining and rejecting several moderate answers to SCQ, van Inwagen defends his own moderate answer that living organisms are composed entities—the *only* composed entities: "The xs compose y if and only if y is an organism and the activity of the xs constitutes the life of y."[32] This view implies that there can exist but two types of spatial entities: mereological atoms (simples) and living organisms (composed of mereological atoms). van Inwagen's view is moderate insofar as it allows that at least *some* composite spatial entities exist (viz., living organisms), *and* it also allows that there can exist spatial entities that fail to be composing parts of other spatial entities (viz., any mereological atom that is not a proper part of a living organism). Though van Inwagen's view counts as moderate, Living Organism is nihilist with respect to nonliving composed spatial entities: Living Organism implies that there exist *no* materially solid *or* scattered *inanimate* spatial entities. And I object to Living Organism because it is thereby too restrictive. Living Organism implies counterintuitively that there exist *no* materially solid three-dimensional subatomic particles, *no* (scattered) hydrogen atoms, *no* (scattered) carbon molecules, and no diamonds, decks of cards, *or* planets. Living Organism also implies counterintuitively, then, that if mereological atoms compose a living human at time t and if that human dies a moment later at time t', there could exist no corpse that would be composed of exactly the same mereological atoms that, a moment earlier, composed a living human.[33,34]

Summary. I conclude that Conjoining *is* a reasonable answer to SCQ. Unlike the Nihilist's and Universalist's answers, Conjoining preserves the commonsense views that there possibly exist at least some spatial entities composed of proper parts, that there possibly exist two disparate spatial entities that do not themselves compose a third, that it is possible that some spatial entity comes into being by composition (i.e., by way of certain spatial entities coming to compose it), that it is possible that some spatial entity ceases to exist by decomposition (i.e., by way of certain spatial entities ceasing to compose it), and that it is possible that there exists a composed spatial entity whose proper parts could exist at a time that the composed entity does not. And, unlike the competing moderate answers addressed, Conjoining obviously allows for the existence of inanimate scattered spatial entities.

Consider now reductivist theories of boundaries and holes cast in terms of spatial directional relations—reductivist theories that are consistent with the directionalist account of materially solid spatial entities and with the existence of both the nonscattered *and* scattered spatial entities that CON allows.

NOTES

1. For analyses of the concepts of a spatial entity, a proper part, nonoverlapping spatial entities, and composition, see D2.1, D3.5, D3.6, and D3.8.

2. See Hestevold, "Conjoining," pp. 371–85. Peter van Inwagen is the philosopher who first referred to this metaphysical problem as "The Special Composition Question." See *Material Beings* (Ithaca, NY: Cornell University Press, 1990), pp. 20, 21–32, and fn. 14 (p. 287). See Verity Harte, *Plato on Parts and Wholes: The Metaphysics of Structure* (Oxford: Clarendon Press, an imprint of Oxford University Press, 2002), pp. 26–32; Harte frames Plato's work on composition in terms of contemporary concerns.

3. Ned Markosian has argued that three-dimensional objects are simples on grounds that no three-dimensional object has proper parts—on grounds that no three-dimensional object can possibly be a proper part of some other three-dimensional object. See Markosian's defense of MaxCon in "Simples." In *Material Beings*, van Inwagen defends the existence of mereological simples, arguing that the only objects that exist are such (*part*-less) "atoms" and living organisms composed of such "atoms." See also Trenton Merricks, *Objects and Persons* (Oxford and New York: Oxford University Press, 2003). Cf. Theodore Sider, *Writing the Book of the World* (Oxford and New York: Oxford University Press, 2011), pp. 79–82.

4. Some would defend MN on grounds that there is no adequate answer to the Special Composition Question. For a lovely explication of this defense, see Amie L. Thomasson, *Ordinary Objects* (Oxford: Oxford University Press, 2007), pp. 126–27; Thomasson, however, does not herself endorse MN. See also Theodore Sider, "Against Parthood," in Karen Bennett and Dean W. Zimmerman, eds., *Oxford Studies*

in Metaphysics, Vol. 8 (Oxford: Oxford University Press, 2013), pp. 237–93. See also Peter Unger, "There are no Ordinary Things," *Synthese* 41 (1979), 117–54.

5. In 1981, I used 'Conjunctivism' and 'CJ' to refer to what is now known as *Universalism*. Chisholm also used 'Conjunctivism' in this sense in his 1987 essay, "Scattered Objects," reprinted in his *On Metaphysics* (1989); see p. 91.

6. Franz Brentano, Nelson Goodman, and Willard Quine have endorsed Universalism. See Brentano, *Psychology from an Empirical Standpoint*, p. 156; Nelson Goodman, *The Structure of Appearance* (Cambridge, MA: Harvard University Press, 1951), pp. 46–47; Willard van Orman Quine, *Word and Object* (Cambridge: The MIT Press, 1960), pp. 91, 97–100, 120–22. For more recent work, see David Lewis, *On the Plurality of Worlds* (Oxford: Blackwell Publishing, 1986), pp. 211–13; Theodore Sider, *Four-Dimensionalism: An Ontology of Persistence and Time* (Oxford: Clarendon Press, an imprint of Oxford University Press, 2001), pp. 121–39; and James van Cleve, "The Moon and Sixpence: A Defense of Mereological Universalism," *Contemporary Debates in Metaphysics*, ed. John Hawthorne, Sider, and Zimmerman (Oxford: Blackwell, 2008), pp. 321–40.

7. See Kathrin Koslicki, *The Structure of Objects* (Oxford and New York: Oxford University Press, 2008), p. 171.

8. Eric T. Olson, *What Are We? A Study in Personal Ontology* (New York: Oxford University Press, 2007), p. 224.

9. One who rejects UN is not committed to rejecting the principle that, for any two objects, there is a whole of which they are both proper parts. Conjoining allows that there exists a "cosmic whole" CW such that every other spatial entity that exists is itself a proper part of CW. For example, Conjoining allows that the first floor and twentieth floor can each be a proper part of a skyscraper and of CW, but Conjoining does not allow that these two floors—separated by eighteen floors in between!—compose an "aggregate" (i.e., "scattered object") that is itself a proper part of both the skyscraper and CW.

10. Roderick M. Chisholm championed ME in *Person and Object*, p. 52. I continue to use substitute 'proper part' for Chisholm's use of 'S-part'. See also Chisholm's earlier formulation and defense of ME in "Parts as Essential to Their Wholes," *The Review of Metaphysics* 26 (1973), 581–603 and "Mereological Essentialism: Some Further Considerations," *The Review of Metaphysics* 28 (1975), 477–84. van Inwagen rejects ME; see *Material Beings*, p. 54.

11. As was argued earlier, there is no reason for the Spatial Directionalist to endorse the existence of substantivalist space, but D5.1 does leave open the possibility that substantivalist space exists and that two spatial entities would be directly conjoined when unoccupied substantivalist space lies between the two.

12. The ice-cube-between-the-tongs-arms example is on a par with my 1981 example involving the dog walking between the World Trade Center Towers; "Conjoining," pp. 378–79. If *no* spatial entity lies between the two towers, then the towers would compose a whole in virtue of being *directly* conjoined. If a Chihuahua walks between the two towers, then, although the towers would cease to be *directly* conjoined, they would nonetheless compose a whole in virtue of being *partially conjoined*—in virtue of each tower's having a proper part directly conjoined with a

proper-part of the other. Per Mereological Essentialism, however, the whole composed of two directly conjoined towers would not be identical with the whole composed of two partially conjoined towers: There would be a "scattered object" that is a proper part of the former but not the latter.

13. van Inwagen defends the existence of simples, noting that their existence is not self-evident. See *Material Beings*, p. 52.

14. Cartwright, "Scattered Objects," pp. 157–58.

15. Chisholm, "Scattered Objects," p. 91. Though both Cartwright and Chisholm endorse scattered objects, they disagree on how the concept of a scattered object should be analyzed. Cartwright claims that an object is scattered if it occupies "a region of space [that] is disconnected" whereas Chisholm analyzes the concept of a scattered object in terms of *direct spatial touching*, a concept that he takes as primitive.

16. van Inwagen, *Material Beings*, pp. 64–65.

17. Hestevold, *Conjoining*, p. 371. Chisholm's definitions appear in *Person and Object*, p. 152. I substitute 'a proper part' for Chisholm's use of 'an S-part.'

18. van Inwagen, *Material Beings*, p. 64.

19. Cf. Koslicki, *The Structure of Objects*, p. 15. With respect to formulating "Classical Extensional Mereology," Koslicki writes that "The single primitive can be chosen to be parthood . . . , overlap, disjointedness or sum; the other notions are definable in terms of whichever one is taken as primitive."

20. In this and the following two sections of chapter 5, I paraphrase three objections that van Inwagen has raised to the "contact" answer to SCQ: "Two objects compose a whole when they are 'in contact' with (i.e. 'touch') each other—when there is no 'gap' between those two objects." If these three objections—the "odd objects," "identity," and "two beginnings" objections—are compelling objections, then they would be compelling objections not only to the "contact" answer to SCQ but to Conjoining as well.

21. For van Inwagen's "odd objects" objection, see his *Material Beings*, p. 35.

22. H1 and H2 would not become *directly* conjoined at time t, but H1 and H2 would become *conjoined* at t: H1 and H2 have proper parts—their right hands—that become *directly* conjoined at t.

23. Cf. van Inwagen, *Material Beings*, p. 36.

24. See van Inwagen, *Material Beings*, p. 36.

25. See Olson, *What Are We?*, p. 225n.

26. See Olson, *What Are We?*, p. 224.

27. See Olson, *What Are We?*, p. 225n.

28. van Inwagen, *Material Beings*, pp. 33–37. See also John W. Carroll and Ned Markosian, *An Introduction to Metaphysics* (Cambridge and New York: Cambridge University Press, 2010), p. 192 and Alyssa Ney, *Metaphysics: An Introduction* (New York: Routledge, an imprint of Taylor & Francis, 2014), pp. 106–7.

29. In this volume's preface, I note that Roderick Chisholm introduced me to the problem of contact in 1976.

30. See van Inwagen, *Material Beings*, pp. 56–58; see also Carroll and Markosian, *An Introduction to Metaphysics*, pp. 192–93 and Ney, *Metaphysics: An Introduction*, p. 107.

31. van Inwagen formulates and rejects two additional nonextreme answers to SCQ—*Cohesion* and *Fusion*. See *Material Beings*, pp. 58–60; see also Ney's formulations, *Metaphysics: An Introduction*, p. 107. Appealing to Ney's formulation, *Cohesion* is the view that a spatial entity is composed of others when the others cohere—when they are fastened in such a way that "they cannot be pulled apart or moved in relation to each other without breaking." And *Fusion* is the view that a spatial entity is composed of others when "the others are joined together such that there is no boundary" *where* the composing entities are joined together. I would argue that Cohesion and Fusion are both too restrictive because they do not allow (or do not obviously allow) that scattered spatial entities exist: A scattered spatial entity is an entity that has at least two nonoverlapping proper parts that do not cohere to one another and are not fused together. Also, a poorly bound paperback book *is* a spatial entity, but Cohesion would disallow this given that pages could be pulled away from its glued spine one page at a time without tearing a page and without damaging the hardened glue. Finally, the concept of *fusion* is no clearer than the concept of *contact* or *fastening*. Exactly, what *are* boundaries—what *is* it that a spatial entity lacks when its composing parts are fused together? If each of two separated materially solid hemispheres has a circular outer boundary that serves as a part of that hemisphere's surface, then where do those boundaries go when the hemispheres are fused together to compose a sphere? *Is* it possible, as Fusion appears to imply, that there exists *no* boundary between a materially solid sphere's left and right hemispheres?

32. van Inwagen, *Material Beings*, p. 91.

33. In an effort to preserve commonsense claims about ordinary inanimate spatial entities, van Inwagen offers paraphrases cast in terms of mereological simples; see *Material Beings*, pp. 98–114. For example, the defender of Living Organism could argue that although, strictly, no books and no shelves exist, to claim that a book is on the shelf is to claim that mereological simples arranged book-wise are located above other mereological simples arranged shelf-wise. And, presumably there are simples arranged corpse-wise even though no corpses exist. Whether all plausible commonsense inanimate-object claims can be similarly paraphrased in terms that are consistent with Living Organism is not obvious. Can "The shelf is composed of more than a thousand carbon atoms," "Some chair is heavier than some table," and "Some bricks are touching each other" be successfully paraphrased as claims that refer to nothing other than mereological simples? See Gabriel Uzquiano, "Plurals and Simples," *The Monist* 87 (2004), 429–51. See also Matthew McGrath, "No Objects, no Problem?," *Australasian Journal of Philosophy* 83 (2005), 470–76. Theodore Sider has noted another potential problem for van Inwagen: Imagine a possible world in which there exists *no* living organism, but in which all spatial entities are "gunky." This would be a possible world in which any spatial entity is such that any one of its parts does itself have parts—a world in which no mereological atoms exist. In a "gunky" world without living organisms and without mereological atoms, commonsense claims about inanimate spatial entities could not be successfully paraphrased as truths about mereological atoms, and there *would* be an adequate answer to SCQ other than van Inwagen's. See "Van Inwagen and the Possibility of Gunk," *Analysis* 53 (1993), 285–89.

34. Committed to the existence of composed spatial entities, but lacking an answer to SCQ, Markosian has defended *Brutal Composition* [BC]: "There is no true, non-trivial, and finitely long answer to SCQ." See Markosian, "Brutal Composition," *Philosophical Studies* 92 (1998), 211–49. BC is a moderate response to SCQ insofar as it allows that some spatial entities compose others, and some do not. But whether composition occurs in a given case is a brute fact. Sider has objected to BC in *Four-Dimensionalism*, pp. 120–32; see Markosian's discussion of Sider's objection in "Brutal Composition," pp. 237–40.

Chapter 6

Does the Road to Nowhere Include Boundaries and Holes?

PROBLEMS INVOLVING BOUNDARIES

Imagine a black and white, materially solid checkerboard 2 cm thick. The checkerboard's parts would include each of the checkerboard's black squares and white squares, multiple square parts composed of two black squares and two white squares each, and the checkerboard's left and right halves. Presumably, the checkerboard's parts would also include its *surface*—its topmost black-and-white-checkered "outer" boundary. Some would insist that the checkerboard's surface would be a three-dimensional part of the checkboard. But how thick, *exactly*, would such a surface be? A quarter-centimeter thick? An eighth-centimeter? A sixty-fourth? Others may argue that the surface is a *non*-three-dimensional part of the checkerboard—a two-dimensional part with no thickness whatsoever. On this view, would it be possible to strip away the two-dimensional surface, leaving behind a *surface-less* checkerboard? Would there also exist twelve one-dimensional parts that are the checkerboard's edges and eight zero-dimensional parts that are *the* corners of the checkerboard? And, if one strips away a single dimensionless corner, how would a dimensionless corner differ from a dimensionless atomic particle? Or from a monad?

Would there be a rectangular part of the checkerboard that would be the "inner" boundary between the left and right halves of the checkerboard? If so, would it be a part of the left half? The right half? Neither? Both? Would that boundary be three-dimensional? Two-dimensional? Similarly, would that rectangular "inner" boundary itself have an "inner" boundary that separates *its* halves? And of which half-boundary would *it* be a part? Would there be an "inner" boundary between a black square and a neighboring white square? If so, would the boundary itself be black or white? Neither? Both?

Would the checkboard have a two-dimensional square "inner" boundary that separates its top and bottom halves? Would that square boundary itself have a zero-dimensional "inner boundary" that would be not only "*the* center" of that square boundary but also "*the* center" of the entire checkerboard? Would it be possible to exchange "*the* center" of one checkerboard with "*the* center" of a second checkerboard, leaving the checkerboards' other parts intact?

Now, instead of imagining a materially solid checkerboard, imagine a standard black and white checkerboard made of marble—a checkerboard that would be a scattered spatial entity that is a swarm of marble molecules. Would this marble checkerboard have a surface—a topmost "outer" boundary? How thick would this surface be, and would it be "gappy"—would *it* be a scattered spatial entity? If the marble checkerboard's "outer" boundary is *not* a scattered spatial entity or is without thickness, then of what would the checkerboard's surface be composed? Would the marble checkerboard have an "inner" boundary that separates its left and right halves? If so, would that "inner" boundary be a scattered or solid spatial entity? Would it be three-dimensional? Of what would it be composed? In what sense, if any, could the marble checkerboard have an "inner" boundary that is "*the* center" of that marble object?

I have defended Brentano's view that the "outer" and "inner" boundaries of solid spatial entities are *non*-three-dimensional parts of those entities and that these parts are dependent particulars—spatial entities that are *essentially* parts of other entities that are not themselves (non-three-dimensional) boundaries. In my defense of this "dependent particulars" view, I was silent (as was Brentano) with respect to boundaries of scattered objects.[1]

I have always had reservations about Brentano's boundaries-are-dependent-particulars view. Here, I suggest an alternative: After exposing problems with two versions of Brentano's view, I formulate a reductivist theory of boundaries that skirts ontological commitment both to a strange class of dependent particulars *and* to regions of substantivalist space. I suggest reducing reasonable claims that apparently imply the existence of ontologically dependent boundaries to claims about the (three-dimensional) "outer surrounding" parts of materially solid spatial entities; and what such parts involve can be explicated in terms of spatial directional relations. Then, by appealing to a directionalist account of a scattered spatial entity's actual location, I explicate the concept of a scattered spatial entity's *outer boundary* in terms of an "outer part" of its actual location. The chapter ends with a sketch on how the Spatial Directionalist could also develop a reductivist/directionalist theory of holes.

ARE BOUNDARIES DEPENDENT PARTICULARS?

Brentano argued that a boundary of a materially solid spatial entity would be a *non*-three-dimensional *dependent particular*—a spatial entity that could not possibly exist without being a part of *something*:

> [A] point depends on something to which it belongs as a boundary. And this is a part of the nature of a point. . . . One cannot say of any specific continuum that its existence is required for the existence of a particular point. . . . For one and the same individual point to exist, what is necessary is not a certain specific continuum, but rather a continuum in general.[2]

According to Brentano, any given *proper* part of a materially solid spatial entity (e.g., a materially solid cube's left half) could exist "independently" as a "free-standing" entity: It could exist without being a part of *that* entity or of any other spatial entity. On the other hand, a two-, one-, or zero-dimensional "outer" or "inner" boundary could not possibly exist unless there exists *some* spatial entity of which that boundary is a part. The two-dimensional square "outer" boundary that is a materially solid cube's topmost surface could not exist without being a part of some materially solid spatial entity or the other; and the same would be true both of the one-dimensional "outer" boundaries that would be the cube's outer edges and the zero-dimensional "inner" boundary that is "*the* center" of the cube.

Chisholm defended one version of Brentano's view, and I defended another. Below, after explicating the former and then the latter, I note concerns specific to each view and then review several reasons to be skeptical of any theory of boundaries that implies them to be *non*-three-dimensional dependent particulars.

The Chisholm/Brentano theory of boundaries.[3] Chisholm offered the following analysis of the concept of a boundary:

> x is a boundary in y =Df x is a constituent of y; and every constituent of x is necessarily such that there is something of which it is a constituent.[4]

Though Chisholm used the word 'constituent' as I have used the word 'part' (not 'proper part') and the word 'part' as I have used the term 'proper part', Chisholm would agree that spatial entities may have two types of parts: (three-dimensional) proper parts and (non-three-dimensional) boundaries; the latter are *dependent*, the former are not. Chisholm captures this feature of boundaries in his second "boundary" axiom, and he captures yet another feature of boundaries in his first:

I have said, following Brentano, that the concept of a boundary is closely related to that of *total* coincidence. I suggest that the relation is this:

A1 For every x, x is a boundary, if and only if, x is possibly such that there is something with which it wholly coincides.

. . .

The following axiom reflects the fact that boundaries are essentially dependent entities:

A2 For every x, y, and z, if x is a boundary in y, and if z is a part of y in which x is not a boundary, then there is a part of y discrete from z in which x is a boundary.

This principle implies that every constituent of every boundary is a constituent of something that is not a boundary. It is thus inconsistent with Suarez' suggestion . . . according to which God could remove just the surface of a three-dimensional object.

Could God preserve any of the boundaries of a thing apart from the thing? . . . [God] couldn't preserve the boundaries except by retaining *some* part [i.e., proper part] of the original thing.[5]

In close accord with Brentano's work and taking as primitive the concept of *total coincidence*, Chisholm claims that the boundaries of two nonoverlapping spatial entities may themselves *wholly coincide*.[6] (Chisholm does not appeal to substantivalist space, but the substantivalist would claim that two nonoverlapping boundaries wholly coincide when they occupy the same place—the same region of space.) For example, Chisholm would argue that, with respect to a materially solid cube composed of left and right half-cubes, (i) the left and right half-cubes do not overlap, (ii) each could exist when no part of the other exists, (iii) a square two-dimensional "inner" boundary would be the inside surface of the left half-cube, (iv) a square two-dimensional "inner" boundary would be the inside surface of the right half-cube, (v) the "inner" boundary that is the left half-cube's inside surface *wholly coincides* with the "inner" boundary that is the right half-cube's inside surface, and (vi) if the cube's left and right halves are separated 25 cm apart, the half-cubes would exist but would cease to compose a cube, and the boundaries that were formerly the half-cubes' inside surfaces would exist but would no longer wholly coincide. By the same token, Chisholm (and Brentano) would argue, one could construct a materially solid cube by pushing the two materially solid half-cubes together such that a square two-dimensional "outer" surface of one half-cube wholly coincides with a square two-dimensional "outer" surface of the other half-cube.[7]

A concern about the Chisholm/Brentano theory of boundaries. If Chisholm and Brentano are right that the parts of a three-dimensional materially solid entity include both three-dimensional proper parts and non-three-dimensional boundaries, then presumably *all* of these parts are themselves *material* spatial entities. And, if Chisholm and Brentano are right that the boundaries of two *non* overlapping spatial entities can *wholly coincide*, then Chisholm and Brentano are committed to the view that two nonoverlapping *material* entities can *wholly coincide*—that it *is* possible that two *material* entities with no parts in common nonetheless have parts that "overlap the same place." This consequence of the Chisholm/Brentano view is, at best, counterintuitive.[8]

On behalf of the Chisholm/Brentano view, one may insist that two boundaries *can* wholly coincide given that, unlike a materially solid spatial entity's proper parts, boundaries would be *non*material parts of physical spatial entities. If, however, boundaries are *non*material parts of *materially* solid entities, then of what would boundaries be composed? And how would it be possible that three-dimensional materially solid *physical* spatial entities are composed of both their physical proper parts *and* their *non*material boundaries? Reluctant to endorse a theory of boundaries that allowed for coinciding parts, I defended a variation of Brentano's view.

Hestevold's Brentano-esque theory of boundaries. Implicitly, I claimed that boundaries are *non*-three-dimensional parts of physical objects that are not *proper* parts (i.e., not three-dimensional parts) of those objects:

z is a boundary of w = z is a part of w, but z is not a proper part of w.[9]

Avoiding the Chisholm/Brentano claim that boundaries are parts of spatial entities that can wholly coincide, I offered two axioms about the relation between parts and proper parts:[10]

B1 For any x and w, if x is a part of w and x is not a proper-part of w, then x is necessarily such that it is part of an object.[11]

B2 For any x and w, if x is a part of w and x is not a proper-part of w, then there is no part of x which is a proper-part of x.

Coupled with my definition of *boundary*, B1 implies that, per Brentano, boundaries would be those parts of spatial entities that are essentially parts of *something*; and B2 implies that boundaries are those parts of spatial entities that do not themselves have proper parts.[12]

In formulating my 1986 view, convinced that nonoverlapping parts of spatial entities could not possibly "occupy the same place" (i.e., could not possibly wholly coincide), I quietly allowed for the (bizarre) possibility that "*surface*-less" spatial entities exist: With respect to a materially solid cube that has a proper part that is its left half and a proper part that is its right half, (i) no two nonoverlapping parts of the cube can wholly coincide; (ii) a square two-dimensional "inner" boundary would be the inside surface of the left half-cube; (iii) that same square two-dimensional "inner" boundary would also be the inside surface of the right half-cube; (iv) the left and right half-cubes would thereby overlap such that if one (complete) half-cube ceased to be *in nihilum*, the other half-cube (given Mereological Essentialism) could not possibly exist; (v) if the (complete) left half-cube is pulled 25 cm away from the remaining proper part of the cube, the left half-cube would have a square two-dimensional "outer" boundary that was formerly the "inner surface" of the right half-cube; (vi) if the left half-cube is pulled 25 cm away from the remaining proper part of the cube, the remaining proper part would be *surface*-less—it would be the right-half-cube-minus-its-two-dimensional-"inner-surface." My 1986 view also quietly allowed that it is impossible to construct a materially solid cube by pushing together two half-cubes with "complete outer surfaces." Rather, only by pushing together a "*surfaced*" spatial entity with a "*surface*-less" spatial entity could one combine two materially solid entities to compose a materially solid whole.

Concerns about Hestevold's Brentano-esque theory of boundaries. My 1986 view *is* implausible. First, any view that implies that a materially solid sphere cannot possibly be composed of (or cannot possibly be divided into) two nonoverlapping hemispheres *is* implausible.[13] (At best, my 1986 view implies that a sphere would be composed of a hemisphere plus a "partially *surface*-less near-hemisphere"—a hemisphere minus its two-dimensional circular surface.) And, the 1986 view is similarly implausible because it implies that it would be impossible to construct a materially solid sphere by pushing two *surfaced* hemispheres together.

Second, my 1986 view is implausible because it allows that *surface*-less materially solid entities exist. And this is a dubious implication: Among the materially solid spatial entities that exist, which ones are *surface*-less and which are *surfaced*? And, how could one tell the difference between *surface*-less and *surfaced* materially solid spatial entities? In our world, what is the ratio of *surfaced* to *surface*-less spatial entities? Does the ratio change?

Third, my 1986 view is implausible because it does not allow for a nonarbitrary explanation for why the two-dimensional circular "inner" boundary in the middle of a materially solid sphere would "remain attached" to one hemisphere rather than the other if the sphere is divided into two proper parts of which only one could be a hemisphere. Suppose that a sphere is sliced apart such that its left hemisphere remains intact: The two-dimensional circular boundary that was the left hemisphere's "inner surface" would now be a part of the left hemisphere's "outer surface;" what remains of the right hemisphere would be a near-hemisphere that is partially *surfaced* given that the right hemisphere would have ceased to exist (per Mereological Essentialism) when it lost its two-dimensional circular "inner surface" to the left hemisphere. The question for which there would seem to be no nonarbitrary answer is this: When the sphere was divided into two of its composing proper parts, in virtue of what did the "inner" boundary "in between" those two parts "remain attached" to the left hemisphere rather than to the right hemisphere? What could have been done differently such that the two-dimensional circular "inner" boundary would have instead remained a part of the intact right hemisphere such that the left hemisphere, by losing its "inner surface," would have been reduced to a partially *surfaced* near-hemisphere that lacks a two-dimensional circular "outer" boundary?

Fourth, my 1986 view involves a tension between axiom B1 and the view's allowing that *surface*-less materially solid entities exist. B1 implies that any given boundary is necessarily a part of *something* at any time that it exists—that it would be impossible to strip away just "*the* zero-dimensional top" from a materially solid sphere or to strip away just the two-dimensional circular boundary that is a part both of the sphere's left and right hemispheres. Presumably, the defender of B1 should claim that it is obviously impossible that a boundary is entirely detachable from all three-dimensional proper parts of some given spatial entity (and from all other spatial entities as well). That a boundary is detachable from all of an entity's proper parts is not *obviously* impossible if one also insists—as the defender of my 1986 view should— that it is *obviously* possible that a materially solid sphere's two-dimensional circular "inner" boundary is detachable from the left hemisphere *and* also detachable from the right hemisphere. If it *is* possible that the circular "inner" boundary can exist when detached from the left hemisphere and if it *is* possible that the circular "inner" boundary can exist when detached from the right hemisphere, then why suppose that it is *impossible* that the circular boundary exists detached from *both* hemispheres and from all other spatial entities? Why suppose, then, that axiom B1 is true—that it is *not* possible that a boundary exists *without* being a part of *something*? (To put this point in terms of conceivability, if a metaphysician *can* conceive that the circular boundary is detachable from the left hemisphere and also detachable from the right hemisphere, then shouldn't that metaphysician also be able to conceive that the boundary *can* be detached from *both* hemispheres, existing independently without being a part of anything else?)

I do not now know why I once believed that a theory of boundaries that allows for *surface*-less spatial entities would be any less implausible than a theory that allows for wholly coinciding spatial entities! Consider now several additional problems common to both the Chisholm/Brentano view and my 1986 view.

Problems with Brentano-esque views. First, if one sides with Brentano's view that boundaries are non-three-dimensional dependent particulars, then of what, *exactly*, would a boundary be composed? If boundaries are *parts* of three-dimensional *material* entities, then presumably they, too, would be material;[14] but this implication is dubious: It is dubious that two nonoverlapping *material* things could *coincide* ("occupy the same place") as Chisholm would have us believe. And, as my 1986 view allows alternatively, it would also be dubious that there could exist *material* entities that are *surface*-less—the result of a *surfaced* entity's losing its three-dimensional surface. Perhaps Brentano's defender will sidestep these concerns by claiming that boundaries are *non*material parts of materially solid entities. This move is also dubious: How could it be that the parts of a *material* entity would include parts that are *non*material? Thus, if one sides with Brentano's view that boundaries are *non*-three-dimensional parts of *material* entities, one may be hard-pressed to explain of what boundaries are themselves composed.

Second, defenders of a Brentano-esque theory of boundaries may be committed to the problematic view that a three-dimensional materially solid physical entity is composed of indefinitely many nonoverlapping zero-dimensional entities; and whether *this* is metaphysically possible is not obvious. If every materially solid three-dimensional entity has indefinitely many two-dimensional boundaries as parts, and if every two-dimensional boundary has indefinitely many one-dimensional boundaries as parts, and if every one-dimensional boundary is a dense array of zero-dimensional boundaries, *would* Brentano's defender thereby be committed to the view that every materially solid entity *is* an *actu infinitum*—an actual infinitude of zero-dimensional spatial entities that compose a three-dimensional material whole? Whether Brentano's defender would be committed to such a view is itself a problem that would need to be addressed. And, if Brentano's defender *is* committed to such a view, then she will owe skeptics a defense. For example, Aristotle claimed that *no actu infinitum* exists:

> When we speak of the potential existence of a statue we mean that there will be an actual statue. It is not so with the infinite. There will not be an actual infinite.[15]

And, Brentano himself was skeptical about the *actu infinitum*; he argued that his view does *not* imply that materially solid entities are composed of infinitely many zero-dimensional parts:

> He can just as well describe [a materially dense spatial entity] as any number of correspondingly small entities he pleases, but he cannot understand it as an infinite number of infinitely small entities.... One can say of a continuum, then,

only that it can be described as being as large a finite number of actual entities as you please, but not as an infinitely large number of actual entities.[16]

The skeptic demands *some* account of how it is that infinitely many zero-dimensional entities can "mingle" or "combine" to compose a three-dimensional spatial entity. For that matter, the skeptic would demand an account of how it is that a *finite* number of *non*-three-dimensional boundaries and three-dimensional proper parts could "mingle or "combine" to form a materially solid whole.

Finally, a shortcoming of both Chisholm's view and my 1986 view is that neither addresses the boundaries of *scattered* spatial entities. Presumably, garden-variety spatial entities such as rocks, bicycles, and trees have surfaces (i.e., "outer" boundaries), but in what sense could such entities have surfaces, if they are, as scientists tell us, scattered spatial entities—scattered swarms of scattered atoms? At best, then, Chisholm's view and my 1986 view are incomplete theories of boundaries: They address the boundaries of materially solid spatial entities but not of scattered spatial entities.

A third Brentano-esque view. If one is convinced both that boundaries are non-three-dimensional spatial entities that cannot coincide and that *surface*-less materially solid entities cannot exist, then one may be tempted by the view that boundaries can come into being *ex nihilo* and cease to be *in nihilum* with the dividing and combining of materially solid entities.[17] On this view, when a materially solid sphere is halved, the original two-dimensional circular boundary would remain the outer boundary of one of the hemispheres and a new two-dimensional circular boundary would come into being *ex nihilo*, becoming the outer boundary of the other hemisphere. The result would be two separated hemispheres that would each have "complete" surfaces. If the two hemispheres are later pushed together, recombined to compose a sphere, then the two-dimensional circular boundary of one hemisphere would cease to be *in nihilum*, resulting in a materially solid sphere.

This theory of boundaries is more troublesome metaphysically than either Chisholm's view or my 1986 view. First, if boundaries have material composition, then, *exactly*, how *does* matter come into being *ex nihilo* and cease to be *in nihilum*? (If boundaries are *non*material entities, then of *what* would these nonmaterial entities be composed, how would *that* nonmaterial substance come into being *ex nihilo* and cease to be *in nihilum*, and how could nonmaterial things serve as the entities that "bound" *material* spatial entities?) When a sphere is halved, in virtue of what would the two-dimensional circular boundary remain "stuck" to one hemisphere rather than the other? Is the coming into being or passing away of a boundary a non-instantaneous event that happens gradually as one hemisphere cleaves away, or meshes, with another? Or, would boundaries come into being and cease to exist instantaneously as, say, two materially solid hemispheres are pulled apart or pushed together?

Given the reasons to be skeptical of Brentano-esque views that imply that boundaries are essentially non-three-dimensional parts of materially solid spatial entities,[18] consider instead a directionalist theory of boundaries.

DIRECTIONALIST BOUNDARIES

Skeptical that the boundaries of three-dimensional materially solid spatial entities are *non*-three-dimensional parts of those entities, I suggest that boundaries are certain "surrounding" *proper* parts of materially solid entities. Given that the concept of a proper part is analyzable in terms of spatial directionalist relations, this reductivist theory of boundaries can be described as a directionalist theory of boundaries.

"Outer" boundaries (surfaces). A three-dimensional materially solid entity's *outer boundary* is one of its *surrounding* three-dimensional proper parts such that no connected part of the entity "lies beyond" that *surrounding part*. I analyze the concept of a *surrounding part* in terms of certain spatial directional relations that (per D3.7) a spatial entity exhibits internally:

> D6.1 x is a surrounding part of w at time t =Df At time t, w is a materially solid spatial entity that is necessarily such that (i) w is composed of proper parts x and y, and (ii) for any spatial directional relation d that y exhibits internally, there is a proper part of x that bears d to y.

Consider again materially solid sphere Q along with a materially solid croissant C; both are represented in figure 6.1. C is composed of A and B, and (per D6.1) A would be a surrounding part of C: For any spatial directional relation that obtains between two of B's proper parts, there will be a proper part of A that bears that relation to B. For example, d_j obtains between two nonoverlapping proper parts of B, and there is a proper part of A (viz., D) that bears d_j to B.

Regarding the materially solid sphere, D6.1 implies that S, G, *and* H are surrounding parts of Q. Q is composed of "outer shell" S and "core" R; and S is composed of G and H. I is a proper part of S that "lies below" R and is composed of a proper part of G and a proper part of H. Given that R's bottom half bears spatial directional relation d_i to R's top half, d_i is among the indefinitely many spatial directional relations that R exhibits internally. In the case of sphere Q, then, S would be a surrounding part of R given that I would be a proper part of S that bears d_i to R. And, G and H would also be surrounding parts given that, respectively, a proper part of G and a proper part

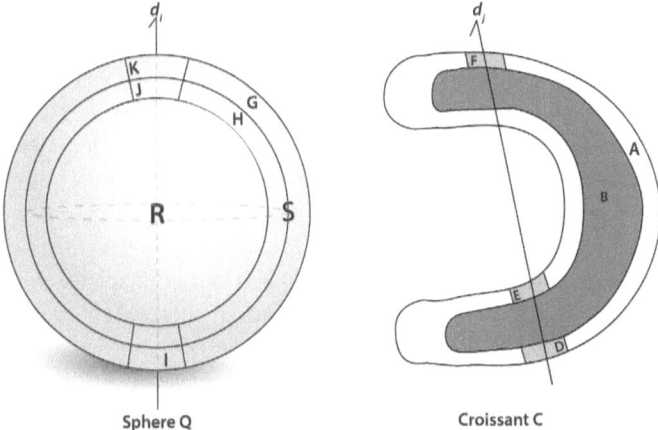

Figure 6.1 Outer Boundaries as Surrounding Parts. *Source:* C. Frantom, J. Rich: White Roche LLC.

of H bear d_i to R. More generally, any surrounding part of Q—any one of Q's proper parts that "completely surrounds" R—will have a proper part that bears to R *any* spatial directional relation that obtains between any two of R's nonoverlapping proper parts.[19]

A materially solid spatial entity has both surrounding parts and *inner parts*:

D6.2 x is an inner part of w at time t =Df At time t, w is a materially solid spatial entity that is necessarily such that (i) w is composed of proper parts x and y, and (ii) y is a surrounding part of w.

One can analyze the concept of a materially solid entity's outer boundary in terms of the spatial directional relations that obtain between the entity's inner and surrounding parts.

A materially solid spatial entity's *outer boundary* (i.e., *surface*) is a surrounding part—a surrounding part such that no (connected) part of the entity that it bounds "lies beyond" that surrounding part:

D6.3 b is the outer boundary (surface) of w at time t =Df At time t, w is necessarily such that (i) w is composed of surrounding part b and inner part y, (ii) for any spatial directional relation d and proper part z of b, if y bears d to z, then there is no proper part of w that is connected with z to which z bears d.

D6.3 implies that both S (composed of surrounding parts G and H) and G itself would count as the outer boundary (surface) of S; and although H would be a surrounding part, D6.3 implies that H would *not* be S's outer boundary. Although S, G, *and* H are all surrounding parts of Q, H does not satisfy the definiens of D6.3: Although R bears d_i to a proper part of H (e.g., J), there *is* a (connected) proper part of S (viz., K) to which that proper part of H bears d_j. Thus, although H is a surrounding part of S, H is not an "outermost" surrounding part.

D6.3 also implies that A is an outer boundary of Croissant C. Inner part B bears d_j to E, which is a proper part of A, but there is no proper part of C that is connected with E to which E bears d_j. Of course, E does bear d_j to F, which *is* a proper part of surrounding part A; but F is not connected with E. Thus, proper part A satisfies the definiens of D6.3 and is thereby an outer boundary (surface) of C.

Consider a potential objection to this directionalist account of outer boundaries (surfaces): "A materially solid object has but one surface: Its outermost part! The directionalist theory implies, however, that such objects have many surfaces. For example, the directionalist theory implies that surrounding parts S *and* G are both outer boundaries (i.e., surfaces) of sphere Q, and the directionalist theory also implies there is a proper part of G *and* a proper part of *that* proper part that would also be outer boundaries of G!" The reply to this objection is that the directionalist theory implies that a given materially solid entity has one surface that *is* many different outer boundaries of the object. Regarding sphere Q, S is an outer boundary that would be thicker and would contain more matter than a thinner outer boundary such as G (and the many proper parts of G that would also be outer boundaries of Q). Each outer boundary of an entity would be identical with a different surrounding part of the materially solid entity, but all of an entity's outer boundaries would be *the same outer boundary* of that entity:[20]

> D6.4 *b* and *b** are the same outer boundary (surface) of *w* at time *t* =Df At time *t*, *b* and *b** are both outer boundaries of *w*, and there is something *b*** that is both an outer boundary of *w* and a proper part of *b* and of *b**.

Regarding sphere Q, two of its different outer boundaries (e.g., S and G) would be *the same outer boundary* in the sense that there is yet another outer boundary (e.g., another surrounding part of Q that is, say, ¾ or ½ or ¼ the width of G) that is a proper part of both S *and* G.

"Inner" boundaries (surfaces). Just as Brentano's view allowed for both "outer" *and* "inner" boundaries that are *non*-three-dimensional, the directionalist theory allows for both outer and inner boundaries that are *three*-dimensional. In short, an *inner boundary* of a materially solid entity would be an outer boundary of a certain proper part of that entity:

> D6.5 *b* is an inner boundary of *w* at time *t* =Df At time *t*, there is exists something *x* that is an inner part of *w*, and *b* is an outer boundary of *x*.

D6.5 captures the view that an inner boundary would be a boundary that "lies below" the outer boundary of a materially solid entity. With respect to sphere Q represented in figure 6.1. One of Q's inner parts would be RJ: The sphere composed of R and J. And, per D6.5, J would be an inner boundary of Q given that J is an outer boundary of RJ.

Can directionalist inner boundaries "lie between" spatial entities? Ordinarily, one would think of an "inner boundary" as something that "lies between" (or "separates" or "divides") composing proper parts of a materially solid spatial entity. The assumption that inner boundaries "lie between" composing parts drew Brentano and Chisholm to the strange view that boundaries are non-three-dimensional *parts* of materially solid spatial entities and that it is possible that two nonoverlapping inner boundaries can *spatially coincide*, that is, can "occupy the same place." This same assumption also motivated my equally strange, if not stranger, 1986 view that allows for the existence of *surface*-less materially solid entities: My 1986 view allows that "halving" a materially solid sphere would produce one hemisphere and one near-hemisphere; that is, one hemisphere and one "partially unbounded" spatial entity that would *be* a hemisphere but for the lack of a two-dimensional circular "outer boundary."

Though the strange implications of Brentano-esque theories of non-three-dimensional boundaries would be a reason to take seriously the directionalist's view that boundaries are conventional *three*-dimensional parts of spatial entities, is there a sense in which the directionalist could allow that a three-dimensional boundary "lies between" the entities that it "separates"? And just how wide would the boundary be between materially solid entities that are the size of North Dakota or the size of a pearl's left hemisphere? To address this question, the directionalist can draw the distinction between boundaries *that lie entirely between* and *common boundaries that lie between* nonoverlapping spatial entities.

A boundary lies entirely between two materially solid spatial entities if it does not overlap either entity that it separates:

D6.6 b is a boundary that lies entirely between x and y at time t =Df At time t, x and y do not overlap, x is entirely connected with b, and b is entirely connected with y.

With respect to a short, materially solid tower of blocks ABC, block B would be the boundary between blocks A and C: A would be entirely connected with B, and B would be entirely connected with C. Similarly, with respect to sphere Q in figure 6.1, H would be the boundary between G and R. D6.6 does indeed allow that a boundary that lies entirely between two entities could be twice the size (or a mere fraction of the size) of either of the other entities.

Two spatial entities have a *common boundary* if they are (overlapping) spatial entities that have a common part that lies entirely between proper parts of the two entities:

D6.7 *b* is a common boundary between *x* and *y* at time *t* =Df At time *t*, (i) *x* is composed of *v* and *b*, (ii) *y* is composed of *w* and *b*, (iii) *b* is a boundary that lies entirely between *v* and *w*.

With respect to materially solid tower of blocks ABC, B would be the common boundary between AB and BC: B is the common proper part of AB and BC that lies entirely between A and C.

Suppose that Mississippi, Alabama, and Georgia are materially solid land masses—respectively, MS, AL, and GA. Per D6.6, AL would itself count as a boundary that lies entirely between MS and GA. But how big is the boundary between AL and MS? If one conceives of AL and MS as two *nonoverlapping* materially solid entities, then there would be *no* land mass that completely lies between these two states. Legislators, law-enforcement agents, and surveyors, however, may give us practical reasons to conceive of AL and MS as entities that *do* have a common boundary—entities that have a long, narrow strip of land as a common part. How narrow would that boundary be? Though one's answer would be arbitrary—1 cm? ½ cm? ¼ cm?—there are practical concerns that suggest that one should stipulate the common boundary to be quite narrow: Which state is responsible for maintaining the segment of a highway that crosses the common boundary? Did the kidnapper transport the victim across "the state line"? What is the narrowest strip of land that surveyors' instruments can measure? Perhaps federal legislators will stipulate that, legally, one should conceive of the common boundary between any two states as ¼ cm wide. One could then conceive of MS as the state composed of MS* and the narrow ¼ cm strip of materially solid land that runs along, and is entirely connected with, the eastern edge of MS*. And, one could conceive of AL as the state composed of AL* and the narrow ¼ cm strip of materially solid land that runs along, and is entirely connected with, the western edge of AL*.

One might object that the directionalist theory has not solved the boundary problem at all—that the problems that plague Brentano-esque theories also plague the directionalist theory:

> You claim that sizeable three-dimensional, materially solid entity AL can be a boundary between MS and GA—two other sizeable three-dimensional, materially solid spatial entities that AL does not overlap. But what about the boundary that lies entirely between MS and AL and the boundary that lies entirely between AL and GA? How big are *those* boundaries? Surely *those* boundaries would be non-three-dimensional parts of the large spatial entity composed of

MS, AL, and GA: One non-three-dimensional boundary would separate MS from AL, and the other would separate AL from GA.

The defender of the directionalist theory of boundaries should remind the critic that MS would be entirely connected with AL and that AL would be entirely connected with GA such that there would be no proper part of the whole composed of MS, AL, and GA that would fail to overlap at least one of these spatial entities. Thus, there could exist *nothing* that would lie entirely between MS and AL or between AL and GA. One could, however, speak of a common boundary between a *proper part* of MS and AL or between a *proper part* of AL and GA, but there would be nothing that lies entirely between *that* common boundary and *that* proper part of MS or AL.

Those who insist that there *do* exist non-three-dimensional boundaries that "separate" such materially solid masses owe both an argument for the view that any materially solid entity is something that, essentially, *has* non-three-dimensional parts *and* an explanation of how it is possible that (i) non-three-dimensional parts can *wholly coincide* ("occupy the same place"), or that (ii) there could exist a *surface*-less materially solid entity produced when a non-three-dimensional entity cleaves to the spatial entity with which the *surface*-less entity had been entirely connected, or that (iii) non-three-dimensional parts come into being *ex nihilo* and cease to be *in nihilum* as materially solid spatial entities are pulled apart and pushed together. One can dodge these challenges by claiming instead that boundaries are three-dimensional proper parts of materially solid spatial entities—parts that are not metaphysically strange. Consider now a directionalist account of boundaries regarding *scattered* spatial entities.

THE BOUNDARIES OF SCATTERED OBJECTS

Newton wrote that "the place of the whole is the same as the sum of the places of the parts."[21] This suggests a theory of boundaries for scattered spatial entities: The boundary of a scattered entity would be something that, in some sense, "surrounds" the smallest "connected place" that the scattered entity occupies—the smallest "connected place" that includes the "scattered places" occupied by the scattered entity's scattered parts.

A scattered spatial entity's possible location is a "large" possible location that "encompasses" the "smaller" possible locations of each of its composing parts—each of its nonscattered nonoverlapping proper parts:

D6.8. Spatial entities x and y overlap at time t =Df At time t, there exists something z such that z is a part of x, and z is a part of y.

D6.9 At time t, spatial entities $x_1, x_2, x_3 \ldots x_n$ are the composing proper parts of scattered spatial entity w =Df At time t, $x_1, x_2, x_3 \ldots x_n$ are such that (i) none is a scattered spatial entity, (ii) each is a proper part of w, (iii) no two overlap, and (iv) any spatial entity that overlaps $x_1, x_2, x_3 \ldots$ or x_n is itself a proper part of w.

If a helium atom were composed of a materially solid nucleus plus two orbiting electrons, then the nucleus and two electrons would be the three composing parts of that scattered atom.

For any scattered atom or any other scattered spatial entity at any given time, there could well be indefinitely many "regions" within which the scattered entity's composing parts are located:

D6.10 At time t, only scattered object w is located within possible location p =Df It is possible that there exists a materially solid entity for which p is the actual location; and at time t, (i) each of the actual locations of the composing parts of w is a subset of p and (ii) no subset of p is the actual location of any spatial entity that is not a proper part of w.

The conception that D6.10 captures is that a scattered spatial entity "lies within" various "connected regions of space": There would be multiple sets of spatial directional relations that would each be the possible location for a single materially solid entity; and the sets of spatial directional relations that are the actual locations of the composing parts of the scattered entity would be subsets of each of these "surrounding connected regions." A scattered atom and its various proper parts could be the only entities located within a "region" that "loosely surrounds" the atom with "a relatively wide margin," but the atom would also be located within a "region" that "tightly surrounds" it. Instead of imagining an atom with a materially solid nucleus, imagine a scattered spatial entity composed of two materially solid spheres 1 cm in diameter and 2 cm apart; and imagine that no other spatial entity exists within 10 cm of either sphere. That scattered object would be the only entity located within indefinitely many different regions. For example, the actual locations of the two spheres would be subsets of a set of spatial directional relations that is the possible location for a 6-cm materially solid cube and of a set that is the possible location for a 7-cm dodecahedron. And, the actual locations of the two spheres would also be subsets of a set of spatial directional relations that is the possible location for an 8-cm sphere, of a set that is the possible location for an 8.1-cm sphere, and of a set that is the possible location for an 8.15-cm sphere.

Regarding the various "regions" within which a given scattered spatial entity is located, some "regions" are smaller than others:

D6.11   Possible location *p* is smaller than possible location *q* =Df *p* and *q* are possible locations for materially solid spatial entities and are necessarily such that the volume of any entity for which *p* is the actual spatial location would be less than the volume of any entity for which *q* is the actual spatial location.

The *actual location* of a scattered spatial entity would be "the smallest region" within which that entity is the only entity located:

D6.12   At time *t*, *p* is actual location of scattered spatial entity *w* =Df At time *t*, only scattered object *w* is located within possible location *p*, and there is no possible location *q* within which *w* is located such that *q* is smaller than *p*.

D6.12 above captures the commonsense view that the location of a scattered entity at a given time would be "the smallest region of space" that "circumscribes and contains" the entity's composing parts.

Earlier, I claimed that the outer boundary of a materially solid entity is any one of its "outermost" surrounding parts. Similarly, I suggest that the *outer boundary* of a scattered object would be any one of the subsets of the scattered entity's actual location that could serve as the actual location for an outer boundary of a materially solid entity that could occupy that location:

D6.13   At time *t*, *b* is the outer boundary [i.e., the surface] of scattered object *w* =Df At time *t*, where *p* is the actual location of scattered spatial entity *w*, *b* is a subset of *p* such that, if *p* were the actual location of a materially solid spatial entity, *b* would be the actual location of one of the materially solid entity's outer boundaries.

An implication of D6.13 is that there would be a subset of the boundary of a scattered object that would be the actual location for a spatial entity that overlaps at least one of the scattered entity's composing parts.

Consider a potential objection to this theory of boundaries regarding scattered objects: "D6.13 implies that a scattered spatial entity that moves 1 meter in any given direction would have a different outer boundary [surface] at each moment that it is in motion, and this is counterintuitive. After all, a diamond is a scattered entity composed of scattered carbon atoms, and the diamond's outer boundary—its surface—would remain the same when the jeweler moves the diamond from her safe to her display case." The directionalist should accept the implication of D6.13 that a moving diamond *would* have a different outer boundary at each time that it is in motion, noting that this is not a bizarre implication.[22] Indeed, *scattered* entities are not entities that have outer boundaries (surfaces) in the same sense that *materially solid* entities have outer boundaries: Though a materially solid entity's outer boundary *is* a part of that entity, there would be no surrounding part of a scattered entity that could serve as *its* outer boundary. The sense in which a *scattered* entity could be bound is the sense in which its composing parts are "confined" to a certain "compact connected location," and that location's "surface" would be an "outermost surrounding sub-location" beyond which no composing part (or part of a composing part) exists. So, when the scattered entity's location changes, its "outermost surrounding sub-location" would change as well. And this would involve nothing metaphysically bizarre.

Consider now how one could appeal to Spatial Directionalism to develop a theory of *holes*.

TOWARD A DIRECTIONALIST THEORY OF HOLES

A doughnut's hole is larger than a needle's hole. The round of Swiss cheese contains more holes than half the round. Such claims imply that *holes* exist. But what *are* holes? One may be tempted to identify holes with "empty spaces"—with regions of substantivalist space unoccupied by any spatial entity. There are reasons to resist this view. First, at the very least, this view should be modified to allow that some holes are filled: A doughnut's hole may be filled with jelly, a needle's hole may be filled with thread, and the hole in one's tooth may be filled with amalgam. Second, if a moving material entity's hole *is* a region of substantivalist space, then the entity would have a different hole at every moment throughout the temporal interval during which it moves. If one moves a doughnut from a baker's rack to a plate, the region of substantivalist space that the doughnut would surround on the rack would not move at all; so, the region of space that the doughnut would surround on the plate would be other than both the space that it would surround on the rack and the space that it would surround midway between the plate and the rack. This is an odd implication given that, ordinarily, one would claim that doughnuts along with their holes *do* move. Finally, I am myself inclined to resist the "empty space" theory of holes because it does require ontological commitment to substantivalist space. There is, then, ample motivation to seek a different theory of holes.

Cautious about ontological commitment to anything other than material objects, David and Stephanie Lewis have argued that holes are material parts of material objects: Every object with a hole has a hole-lining—the part of the object that "surrounds the hole"—and the object's hole-lining *is* the object's hole.[23] On this view, a single materially solid doughnut would have indefinitely many (concentric) hole-linings, but these hole-linings *are* a single hole; a round of Swiss cheese would have indefinitely many *different* holes each of which can be identified with indefinitely many different hole-linings.

The Lewises' hole-lining theory is not without counterintuitive implications. First, one may complain that the Lewises' view implies that holes are composed of matter—the *same* matter that composes the object in which the hole is found. Ordinarily, one would claim that the wedding ring's hole is not composed of gold and that the holes in Swiss cheese are not composed of cheese; rather one would claim that these holes are "absences" of, respectively, gold and cheese. If, however, the Lewises are right, then the tire's hole *is* (a hole-lining that is) composed of gold, and the holes in a round of Swiss cheese *are* (hole-linings that are) composed of cheese.

Second, the Lewises' view involves ambiguity regarding the *inside* of a hole. If a Dutch boy is told to stick his finger *in* the hole of a leaking earthen dike, where should the boy stick his finger? The Lewises' view is that the dike's hole *is* a hole-lining, which would be a proper part of the earthen dike—not "an opening" through which water spurts. So, if the boy sticks his finger in "the opening" through which water spurts, then, according to the Lewises' view, the boy will not have done what he was told to do. To comply with the directive, the boy should not plug the spurting water with his finger but should instead poke his finger into the earth that surrounds the spurting water. I take no stand on whether either of these objections (or others) is a decisive objection to the Lewises' hole-lining theory.

Those who do reject both the holes-are-empty-regions-of-space and holes-are-hole-linings views may side instead with Robert Casati and Achille C. Varzi who claim that holes are *"immaterial bodies,* 'growing', like negative mushrooms, at the surfaces of material bodies."[24] I do not myself understand the concept of an "immaterial" entity that has the capacity to "grow" parasitically at a physical object's surface. Unless such a view can be made much less mysterious, one should be reluctant to expand one's ontology to include *immaterial holes*.

Developing a full-blown theory of holes rests well beyond the scope of this book. Such an undertaking would involve a much more thorough evaluation of the three theories of holes thus far addressed and it would involve the formulation of a theory broad enough to include not only doughnut holes and cavities, but holes dug by dogs, dents in fenders, a soap bubble's hollow interior, fissures, valleys, canyons, and tunnels. And, the theory should distinguish between holes "filled with nothing" (e.g., doughnuts and bubble-shaped entities in a vacuum) and filled holes (e.g., holes filled with jelly, with a finger, or simply with air molecules). I suggest, however, that, when entertaining again the views that identify holes with regions of substantivalist space, material hole-linings, or "immaterial bodies," one should also entertain a reductivist theory of holes formulated in terms of spatial directional relations.

This chapter closes with analyses that point the way toward the development of a directionalist theory of holes.

Unfilled holes within materially solid objects. Imagine a materially solid doughnut, wedding ring, and soap bubble located in a vacuum such that *nothing*—no jelly, no finger, no air molecules—fills the doughnut's, ring's, or bubble's hole. Any such spatial entity that "contains" or "hosts" an *unfilled* hole is an entity that has at least one proper part that is a scattered spatial entity—a proper part composed of "entirely separate" parts that are directly conjoined. In a vacuum, there is a proper part of the doughnut's upper-right quarter that is *directly conjoined* with a proper part of the doughnut's lower-left corner, and those two proper parts are not connected with one another given that, between them, there is an empty space—a hole. Given that similar observations can be made with respect to the ring and the bubble, such observations may inspire a hole reductivist to identify holes with empty spaces.

Peter Simons is skeptical that a spatial relationalist can provide a plausible relationalist account of an empty space:

> [T]here is the opposition between substantivalist and relationist accounts of space and time. My sympathies have always been on the ontologically sparser relationist side of this dispute. The main difficulty for relationism has always been making sense of the notion of empty spaces.... This problem is reduced if not eliminated completely, once we recognize that as a matter of fact there are no empty spaces ...: that any location in space ... is the location of something.[25]

Instead of claiming, as Simons suggests, that every relationalist location is occupied, the Spatial Directionalist could claim instead that there could exist *empty locations*—*possible* directionalist/relationalist locations are not *actual* locations. And this would allow the Spatial Directionalist to identify holes with *empty* (directionalist/relationalist) *locations*:

> D6.14 At time t, h is an unfilled hole within materially solid w =Df At time t, (i) z is a scattered spatial entity that is a proper part of w, (ii) z is composed of x and y, (iii) h a possible location, (iv) h, x, and y are such that if a materially solid spatial entity p occupies h at time t, there would exist both a materially solid spatial entity composed of x and p and a materially solid spatial entity composed of y and p, and (v) neither h nor a subset of h is the actual location of any spatial entity.

Consider now a directionalist/relationalist account of *filled* holes.

Filled holes within materially solid objects. Assume that an adequate theory of holes should allow that *filled holes* could exist—that there could exist a materially solid doughnut with a hole filled with materially solid jelly and that there could exist a materially solid balloon filled with a materially solid liquid. Assume as well that the Lewises are right regarding filled holes:

> If you were right that a hole made of cheese could be entirely filled with the same kind of cheese, . . . there would be no such thing as cheese without holes in it. But you are wrong. A hole is a hole not just by virtue of its own shape but also by virtue of the way it contrasts with the matter inside it and around it.[26]

For the purposes of formulating a directionalist/relationalist analysis of *filled* holes, assume with the Lewises that something is a filled hole only if it is filled with a substance other than the substance of the spatial entity that "surrounds" it. Thus, it would be false that a materially solid, nonscattered, emerald-cut diamond-like stone could be a rectangular diamond-like stone with a rectangular hole that is filled with a solid rectangular diamond-like stone. The Spatial Directionalist could allow that a filled hole is a possible location that is the actual location of a spatial entity composed of a substance that contrasts with the substance of the "surrounding" entity:

> D6.15 At time t, h is a filled hole within materially solid w =Df At time t, (i) h is the actual spatial location of something z, and there is a materially solid object v composed of w and z, (ii) w and z are not composed of the same substance, (iii) x is a proper part of w, and there is a materially solid proper part of v that is composed of x and z, (iv) y is a proper part of w, and there is a materially solid proper part of v that is composed of y and z, (v) no proper part of x is connected with any proper part of y.

With respect to a materially solid spatial entity that is "host" to a filled hole, there will be at least two nonoverlapping proper parts of the entity that are "entirely separated" in the sense that neither is entirely nor partially connected with the other, and the actual location of the hole's "filling" is between those two "entirely separated" parts of the "host" entity.

Though the above analyses of unfilled and filled holes suggest that a directionalist/relationalist reductivist theory of holes is worth investigating, I take no stand here on whether such a theory would be preferable to its competitors.

SUMMARY

By invoking Spatial Directionalism, I have used spatial directional relations to formulate the Directionalist Theory of Space. And, DTS *is* a relationalist theory of space that allows for directionalist/relationalist accounts of a spatial entity's *location* and the world's dimensionality, allowing that, for all we know, there exist more than three spatial dimensions. Under the assumption that (absolute) directional relations *do* obtain among spatial entities, I have defended DTS against standard substantivalist objections involving uniform motion, spatial orientation, uniform expansion, and absolute motion; and the directionalist/relationalist replies to these objections preserve the substantivalist spatial intuitions that give the objections their bite. Given the virtues of DTS and given that DTS is ontologically leaner that the substantivalist theory of space, should one not conclude that it is at least reasonable to endorse DTS?

Drawing such a conclusion would be premature: With respect to *locations*, there are substantivalists who agree that regions of substantivalist *space* do *not* exist but insist instead that there *do* exist regions of substantivalist *spacetime*. If *these* substantivalists are right, then one would be left with an ontology that *is space*-less but not *spacetime*-less—an ontological commitment that *would* distress the committed Leibnizian. There is hope, however, that the Spatial Directionalist can continue the trip to no*where* by pointing the way toward a relationalist theory of spacetime cast in terms *spatiotemporal* directional relations. Though resolving the disagreement between *spatiotemporal* substantivalists and *spatiotemporal relati*onalists rests well beyond the scope of this book, I identify in the final chapter some of the issues that metaphysicians, physicists, and philosophers of physics should address to forge a resolution of the disagreement between substantivalists and relationalists.

NOTES

1. Hestevold, "Boundaries, Surfaces, and Continuous Wholes," 235–45. The core of this essay was extracted from my dissertation, "A Metaphysical Study of Aggregates and Continuous Wholes." In the dissertation, I did formulate a theory of boundaries for scattered spatial entities (i.e., *aggregates*), identifying the boundaries of scattered entities with certain *places*—certain regions of substantivalist space.

2. Brentano, *The Theory of Categories*, p. 20.

3. This section is an explication of the view developed by Chisholm in "Boundaries," pp. 83–89.

4. Chisholm, "Boundaries," p. 85.

5. Chisholm, "Boundaries," pp. 85–86.

6. Cf. Newton, *De Gravitatione* in *Philosophical Writings*, ed. Andrew Janiak, Cambridge Texts in the History of Philosophy, series eds. Karl Ameriks and Desmond M. Clarke (Cambridge, New York, Melbourne, Madrid, Cape Town, Singapore, São Paulo: Cambridge University Press, 2004), p. 22, https://www.hrstud.unizg.hr/_download/repository/Newton,_Philosophical_Writings.pdf. After claiming that "space can be distinguished into parts whose common boundaries we usually call surfaces,"

Newton concludes that "surfaces do not have depth, nor lines breadth, nor points dimension, unless you say that coterminous spaces penetrate each other as far as the depth of the surface between them, namely what I have said to be the boundary of both or the common limit; and the same applies to lines and points."

7. In "Boundaries," Chisholm uses his definition of boundary to define the concepts of entities that are *0-, 1-, 2-,* or *3- dimensional,* and he then closed his essay by explicating the concept of continuity in terms of entities that are *in contact* with one another.

8. Cf. Newton, *De Gravitatione,* p. 13: "[A] body so completely fills [space] that it wholly excludes other things of the same kind or other bodies, as if it were an impenetrable being."

9. Hestevold, "Boundaries, Surfaces, and Continuous Wholes," p. 239.

10. Hestevold, "Boundaries, Surfaces, and Continuous Wholes," pp. 238, 239.

11. Compare B1 with Koslicki's formulation of "Priority of Wholes over Parts": "Objects that are part of a whole are *essentially* part of a whole." *The Structure of Objects,* p. 113. See Harte, *Plato on Parts and Wholes,* p. 279; noting that "[t]he claim that parts are structure-laden is thus the claim that there is some sort of metaphysical dependence of the parts on the whole," Harte observes that "how much dependence should be understood would then need to be made clear."

12. In "Boundaries, Surfaces, and Continuous Wholes," I used my definition of *boundary* to define the concepts of zero-, one-, and two-dimensional *inner* and *outer boundaries* and defined the concept of a *continuous whole* (i.e., materially dense spatial entity) in terms of an entity's *surface* (i.e., outer boundary). I then analyzed the concept of an *aggregate* (i.e., a scattered spatial entity) in terms of spatial entities composed of continuous wholes that are not continuous with one another.

13. Cf. Arntzenius, *Space, Time, and Stuff,* p. 128: "*Prima facie,* I find it hard to believe that there really are distinctions between topologically open and topologically closed regions in nature."

14. Cf. Koslicki, *The Structure of Objects,* p. 176: "That the relation between a structured whole and its *material* components is that of parthood I take to be fairly obvious and uncontroversial."

15. Aristotle, *Physics,* III.6.206a18-206b15, trans. R. P. Hardie and R. K. Gayle, *The Basic Works of Aristotle,* ed. Richard McKeon (New York: Random House, 1941).

16. Brentano, *Psychology from an Empirical Standpoint,* p. 354. Though Aristotle and Brentano were skeptical about the *actu infinitum,* Leibniz was not. See Leibniz's *Monadology,* p. 65: "[A]ny portion of matter is not only infinitely divisible . . . but also actually subdivided *ad infinitum.*"

17. Claiming that this view has been attributed Bernard Bolzano, Chisholm cites Bolzano's *Paradoxes of the Infinite* (1851). See Chisholm, "Boundaries," p. 85.

18. Frank Arntzenius would likely claim cautiously that these reasons to be skeptical of Brentano-esque theories of boundaries do not constitute decisive objections to such theories. See Arntzenius, *Space, Time, and Stuff,* pp. 133–34.

19. Replacing clause (ii) of D6.1 with the following would be a mistake: "for any spatial directional relation d that y exhibits internally, x bears d to y." Strictly, S does

not itself bear a spatial directional relation to R: Because S "completely surrounds" R, S is not "in its entirety" separated apart from R. See the discussion of figure 2.3 in chapter 2.

20. This reply is not unlike the Lewises' reply to a similar objection to their hole-lining theory of holes. See David and Stephanie Lewis, "Holes," p. 209.

21. Def.8, Schol.III, *Newton's Mathematical Principles*, p. 7.

22. Those who endorse Mereological Essentialism would readily agree that a diamond in motion would likely have different "outer parts" at each successive time during its period of movement: Whatever counts as a diamond's outer part at a time t would likely undergo mereological change (i.e., would lose or acquire proper parts) between t and any time t' that occurs "very soon" after t.

23. David and Stephanie Lewis, "Holes," pp. 206–12.

24. Casati and Varzi, *Holes and Other Superficialities*, p. 34.

25. Simons, "Where It's At: Modes of Occupation and Kinds of Occupant," pp. 59–60. Cf. Maudlin, "Buckets of Water and Waves of Space," 185. Maudlin explains how a relationalist can dispense with ontological commitment to a *vacuum*: "To say a vacuum exists midway between A and B [that are four meters apart] is just to say that nothing at all exists which has the distance relations of being two meters from each of them." Maudlin concludes that the "relationalist is no more ontologically committed to the vacuum as an entity than one who accepts that there is nothing in the fridge is ontologically committed to the existence of nothing."

26. David and Stephanie Lewis, "Holes," p. 210.

Chapter 7

Is Modern Physics a Roadblock to Going Nowhere?

In earlier chapters, I formulated the Directional Theory of Space, offered replies to standard substantivalist objections to relationalist theories, and then developed the implications of the DTS for metaphysical problems involving the composition of objects and objects' boundaries and holes. Some may be tempted to dismiss this philosophical investigation altogether on grounds that it is quaint—that there is no longer a need for theories of *space* and *time* given that physicists and astronomers assure us that whatever is located is *not* located in *space* and also in *time* but is instead located in *spacetime*. Advocates of *spacetime* may insist that arguing whether one should accept a substantivalist or relationalist theory of *space* or *time* amounts to little more than idle philosophizing given that *our* world *is* a world of physical entities with *spacetime* locations.

In this chapter, I explain why the philosophical investigation of DTS is *not* irrelevant to contemporary concerns about spacetime: The disagreement between substantivalists and relationalists regarding *space* is analogous with the disagreement between substantivalists and relationalists regarding *spacetime*; and there is promise that an appeal to *spatiotemporal* (instead of *spatial*) directional relations could strengthen the relationalists' case. In the first subsection, there is a formulation of Spatiotemporal Substantivalism and a brief explanation of why the favored interpretation of General Relativity includes commitment to substantivalist spacetime. In the second subsection, there is discussion of the implications of Spatiotemporal Substantivalism for the theories of space addressed earlier in this volume: *Spatial* Substantivalism, *Spatial* Relationalism, and the Directional Theory of *Space*. The third and fourth subsections are devoted to explicating *Spatiotemporal* Directionalism and Directionalist *Spatiotemporal* Relationalism—the analogs of Spatial Directionalism and the Directional Theory of Space. If, *ceteris paribus*, ontological simplicity tips the scales from *Spatial* Substantivalism toward DTS, then ontological simplicity would also tip the scales from Spatiotemporal Substantivalism toward Directionalist Spatiotemporal Relationalism *unless* there is a compelling case for countenancing substantivalist spacetime. I sketch the physicist's case on behalf of substantivalist spacetime in the fifth subsection; and I sketch a metaphysician's case against it in the sixth. After evaluating the case against substantivalist spacetime in the seventh subsection, I close the chapter with a review of what is needed from physicists and philosophers of physics to resolve the dispute between substantivalists and relationalists regarding spacetime and why the directionalist's view is philosophically significant even if substantivalists prevail.

SPACETIME, SUBSTANTIVALISM, AND THE FAVORED INTERPRETATIONS OF SPECIAL AND GENERAL RELATIVITY

Michael Friedman writes that "Einstein's central philosophical motivation in developing the general theory [of relativity] was his desire to implement a fully relational, or relativistic, conception of motion, thereby vindicating Leibniz's relationalist attitude toward space and time over Newton's absolutist attitude." Friedman adds, however, that "[w]e now know that general relativity, despite its name, does not in fact fulfill these relationalist ambitions."[1] Though the differences between Special Relativity [SRL] and General Relativity [GRL] are not immediately relevant to the investigation of directionalism at hand, what *is* relevant is that, contrary to Einstein's hope, the favored interpretations of SRL and GRL [Relativity] imply that substantivalist *spacetime* exists. That is, the favored interpretations of Relativity imply that there exist *no* regions of substantivalist *space* and *no* substantivalist *times*, but there do exist regions of a substantivalist "four-dimensional manifold of spacetime points"[2]—*spacetime* regions that *are* the places occupied by things that have locations:

> STS It is false that regions of *space* exist, and it is false that any *time* exists, but there do exist regions of *spacetime*.[3]

Spatiotemporal Substantivalism implies that our *non*spatial, *non*temporal world is instead a *spatiotemporal* world and that *located* entities (e.g., events, objects, boundaries, mereological simples, monads, souls) would be *spatiotemporal entities*—entities that would occupy at least one spacetime point.[4]

As philosophers Lawrence Sklar and Theodore Sider and physicist Brian Greene make clear, favored interpretations of Relativity imply STS given that the favored interpretation implies that *spacetime* points exist and compose the whole of *spacetime*:[5]

> For special relativity the appropriate structure is that of Minkowski spacetime. . . . Now as Minkowski spacetime is the container, it is events that are the contents. The ideal unextended material entity is the concrete event, and the extended material entity is the "world-history," consisting of the "history" of an object taken as the set of happenings that constitute its life. But the rejection of objects for histories, and of space and time for Minkowski spacetime, still leaves much unchanged from this point of view. Spacetime has its independent existence and structure, as space and time did before. So, for example, a spacetime totally empty of all events is perfectly intelligible in this view.[6]
>
> Minkowski spacetime . . . consists of a four-dimensional manifold of spacetime points that contains all of what happens in what we normally call the past, present, and future.[7]
>
> Absolute space does not exist. Absolute time does not exist. But according to special relativity, absolute spacetime does exist.
>
> . . .
>
> [T]he most straightforward reading of Einstein and his general relativity is that . . . *spacetime is a something*.[8]

By implying that the whole of spacetime exists, as the passages above suggest, the favored interpretation of Relativity also implies the Spacetime Parity Thesis—the view that every spacetime region is on an "ontological par" with every other spacetime region:

> STPT For any regions of spacetime s and s', it is not true that the state of the world at s or s' uniquely reflects "the way things really are"; rather, the way things *really* are includes both the way things are at s and the way things are at s'.

To claim that all spacetime regions are on an ontological par is to claim that no spacetime region is more or less real than any other region—that the spacetime region occupied by Socrates is as real as the spacetime region occupied by your last-born great-great-grandchild.

According to STPT, then, there would be no "flow" or "passage" of spacetime such that "future" spacetime regions *become* present and then *become* past. Put another way, *Static Spacetime* is a corollary of STPT:

> SST Spacetime is static in the sense that no spatiotemporal entity undergoes "spatiotemporal becoming," changing from being future, to being present,

and then to being past: Any spatiotemporal entity that occupies a particular region of spacetime *tenselessly* occupies that region.[9]

Imagine that Bailey is a person who dies at age eighty. If there is no "flow" or "passage" of spacetime, then Bailey's death (tenselessly and eternally) occurs in a spacetime region that is *after* the spacetime region where Bailey's fortieth birthday (tenselessly and eternally) occurs, and Bailey's birth (tenselessly and eternally) occurs *before* the spacetime region in which Bailey's fortieth birthday (tenselessly and eternally) occurs. In short, according to the STPT/SST view, distinct regions of spacetime are equally real parts of the whole of spacetime in much the same way that the numbers 0, 40, and 80 are engraved in equally real but distinct locations on a meterstick. (Given that the favored interpretation of Relativity implies the Spacetime Parity Thesis and thereby Static Spacetime, the favored interpretation is inconsistent with theories of Transient Time—with theories that imply that "the future" *becomes* "the present" and then "the past."[10])

Under the assumption that the favored interpretation of Relativity implies the Spacetime Parity Thesis and Static Time, the favored interpretation presumably implies as well that persisting spatiotemporal entities persist via *Spacetime Perdurance*:[11]

> SPER The concept of persistence with respect to spatiotemporal entities is to be analyzed in terms of an entity's being composed of (spatiotemporal) parts such that no one of its (spatiotemporal) parts occupies more than a single region of spacetime; and it is possible that at least one spatiotemporal entity persists.

SPER is the view that, just as a whole loaf of bread "persists" across space in the sense that it is a spatially extended object composed of distinct slices that occupy successive locations, a spatiotemporal entity persists in the sense that it is composed of successive spatiotemporal parts such that each part occupies a successive subregion of that spacetime region that is occupied by the whole persisting spatiotemporal entity.

Consider an argument for the view that the Spacetime Parity Thesis requires a Perdurantist account of persistence. Assume that Bailey is born in New York in 1917, celebrates her fortieth birthday in Chicago in 1957, and dies in Seattle in 1997. If STPT is correct, then Bailey persists from 1917 until 1997, but only in the sense that she *spacetime perdures*. If STPT is correct, then New York in 1917, Chicago in 1957, and Seattle in 1997 are all equally real: the 1917 spatiotemporal part of New York (tenselessly and eternally) occupies one ontologically robust region of spacetime, the 1957 spatiotemporal part of Chicago (tenselessly and eternally) occupies a different but equally robust region of spacetime, and the 1997 spatiotemporal part of Seattle (tenselessly and eternally) occupies still another *real* region of spacetime. If, as STPT implies, these three spacetime locations are distinct but equally real, then, no person who is (tenselessly and eternally) being born in 1917 New York can be identical with a person who is (tenselessly and eternally) celebrating a fortieth birthday in 1957 Chicago; and no person celebrating a fortieth birthday in 1957 Chicago can be identical with a person who is (tenselessly and eternally) dying in 1997 Seattle. Just as a whole loaf of bread is composed of distinct and equally real slices, STPT implies that Baily would be a *whole* person who extends across spacetime from 1917 New York to 1997 Seattle and that Bailey would be composed of distinct and equally real spatiotemporal parts. Bailey's temporal parts would include the part that is (tenselessly and eternally) being born in 1917 New York, the part that is (tenselessly and eternally) celebrating in 1957 Chicago, and the part that is (tenselessly and eternally) dying in Seattle.

STPT, then, is inconsistent with *Endurantist* theories of persistence.[12] The Endurantist claims that an entity persists only if that entity *wholly* (or *intactly*) exists at successive temporal locations—only if the entity at one temporal location *is* identical with an entity that exists at a later temporal location. One who couples Endurantist persistence with STPT would be committed to the implausible view that the robust 7-pound newborn Bailey that (tenselessly and eternally) robustly exists in 1917 New York is *identical* both with the 125-pound Bailey that robustly exists in 1957 Chicago (tenselessly and eternally celebrating her fortieth birthday) and with the Bailey who robustly exists across the country in 1997 Seattle where she is (tenselessly and eternally) dying.

The favored interpretation of Relativity: a summary. The favored interpretation of Relativity implies Spatiotemporal Substantivalism, the Spacetime Parity Thesis, Static Spacetime, and Spacetime Perdurance. For a clearer conception of what this view involves, conceive of spacetime as a dense, four-dimensional array of ontologically robust, "static" spacetime *points*–points that are all equally real and that do not come into being and pass away with "spatiotemporal passage." And, some regions of this array of spacetime points are occupied by persisting spatiotemporal entities—by four-dimensionally extended entities composed of spatiotemporal parts such that none of a persisting entity's spatiotemporal parts occupies more than a single region of spacetime.[13]

In *Slaughterhouse Five*, novelist Kurt Vonnegut describes what the perception of persisting four-dimensional spatiotemporal entities extended across spacetime would be like. Vonnegut's protagonist, Billy Pilgrim, visits Tralfamadore—a planet located outside of *our* four-dimensional spacetime manifold—and Billy Pilgrim subsequently reports that the Tralfamadorians view spacetime's "moments" (i.e., "planes" or "slices" of spacetime points) just as we "can look at a stretch of the Rocky Mountains." From their perspective outside the spacetime manifold, the Tralfamadorians can at once perceive the successive distinct spatiotemporal parts of a persisting entity extended across spacetime just as, from a distance, one can at once perceive the successive distinct mountains that constitute a segment of the Rocky Mountains extended across the Colorado horizon. Vonnegut offers lovely descriptions of what, from a perspective outside the manifold, the perception of persisting (four-dimensional) stars and humans would be like:

> Billy Pilgrim says that the Universe does not look like a lot of bright little dots to the creatures from Tralfamadore. The creatures can see where each star has been and where it is going, so that the heavens are filled with rarefied, luminous spaghetti. And Tralfamadorians don't see human beings as two-legged creatures, either. They see them as great millipedes—"with babies' legs at one end and old people's legs at the other," says Billy Pilgrim.[14]

In summary, then, the endorsement of the favored interpretation of Relativity appears to commit one to the following theses:

- *Spatiotemporal Substantivalism*: There exist regions of *spacetime*.
- *The Spacetime Parity Thesis*: All regions of spacetime are equally real: *No* spacetime region is "ontologically privileged," and *no* spacetime region corresponds with "the way that world really *is*."
- *Static Spacetime*: There is *no* "temporal becoming": It is false that future spacetime regions or spatiotemporal entities will *become* present and *then* past. Rather, any spacetime region or spatiotemporal entity that exists or occurs does so tenselessly and eternally.
- *Spacetime Perdurance*: A persisting spatiotemporal entity persists in the sense that it is composed of distinct successive spatiotemporal parts occupying successive regions of spacetime—regions that are themselves subregions of the spacetime region occupied by the whole persisting (four-dimensional) spatiotemporal entity.

In earlier chapters, I suggested a modest case against *Spatial* Substantivalism: "It is reasonable to reject *Spatial* Substantivalism given that Spatial Relationalism is ontologically simpler and that, when coupled with spatial *directional* relations, Spatial Relationalism can preserve the Spatial Substantivalist's plausible intuitions about uniform movement, spatial orientation, uniform expansion, and motion-caused forces." What should one conclude about the case against *Spatial* Substantivalism if findings of modern physics *do* suggest compelling evidence for the favored interpretation of Relativity and thereby for *Spatiotemporal* Substantivalism? And, if the evidence favors *Spatiotemporal* Substantivalism, then what, exactly, would be the implications for *Spatial* Relationalism and for the earlier-addressed substantivalist/relationalist dispute regarding *space*? Are findings of modern physics a roadblock to going *no*where—a roadblock to endorsing a *location*-less ontology?

THE FAVORED INTERPRETATIONS OF RELATIVITY VIS-À-VIS THEORIES OF SPACE

If modern physics *does* provide compelling evidence for the favored interpretation of Relativity, then there *would* thereby be compelling evidence for Spatiotemporal Substantivalism. And, one could then appeal to modern physics to strengthen the cases against *Spatial* Substantivalism, *Spatial* Relationalism, *and* the Directional Theory of *Space*:

- If there is compelling evidence that *Spatiotemporal* Substantivalism is correct, there would thereby be compelling evidence that *Spatial* Substantivalism is false: The former implies that regions of substantivalist *spacetime* exist and that regions of substantivalist *space* do *not*.
- If there is compelling evidence that Spatiotemporal Substantivalism is correct, then there would thereby be compelling evidence that *space*-implying and *location*-implying claims should be construed as (substantivalist) *spacetime*-implying claims. And, if there is compelling evidence that *space*-implying and *location*-implying claims should be construed as (substantivalist) *spacetime*-implying claims, then there is compelling evidence that *space*-implying and *location*-implying claims *cannot* be reformulated as *non*-space-implying claims involving *spatial* relations. Thus, the evidence for Spatiotemporal Substantivalism is evidence that Spatial Relationalism is false.
- For the same reason, if there is compelling evidence that Spatiotemporal Substantivalism is correct, there would thereby be compelling evidence that the Directionalist Theory of Space is false: Evidence for Spatiotemporal Substantivalism is evidence that *space*-implying and *location*-implying claims should be reformulated in terms of substantivalist *spacetime* and not in terms of spatial directional relations as DTS implies.

If findings of modern physics *do* provide a compelling case for the favored interpretation of Relativity and thereby for Spatiotemporal Substantivalism, and if the case for substantivalist *spacetime* is a case against substantivalist *space*, this would be but a consolation prize for the committed relationalist who resists *all* ontological commitment to substantivalist *places* or *locations*.

SPATIOTEMPORAL RELATIONALISM

Oliver Pooley crisply characterizes the dispute between Spatiotemporal Substantivalists and Spatiotemporal Relationalists:

> Substantivalists maintain that a complete catalog of the fundamental objects in the universe lists, in additional to the elementary constituents of material entities, the basic parts of spacetime. Relationalists maintain that spacetime does not enjoy a basic, nonderivative existence. According to the relationalist, claims apparently about spacetime itself are ultimately to be understood as claims about material entities and the possible patterns of spatiotemporal relations that they can instantiate.[15]

So, if a relationalist continues to find regions of four-dimensionally extended *spacetime* as philosophically repugnant as regions of three-dimensionally extended *space* (that exist at successive *times*), then the relationalist must press the case for Spatiotemporal Relationalism:[16]

> STR It is false that *spacetime* or regions of *spacetime* exist; reasonable claims that appear to imply that *spacetime* or regions of *spacetime* exist can be reformulated as claims involving spatiotemporal entities and spatiotemporal relations that clearly do not imply that *spacetime* or regions of *spacetime* exist.

SPATIOTEMPORAL DIRECTIONALISM

If the relationalist who has been heretofore sympathetic with the Directionalist Theory of *Space* becomes convinced that our world *is* a *spatiotemporal* world (and not a world that is, as Leibniz and Newton believed, both spatial *and* temporal), then the relationalist may be tempted to develop a *directionalist* theory of *spacetime* formulated in terms of *spatiotemporal* directional relations.

A commitment to spatiotemporal directional relations would be a commitment to Spacetime Directionalism—the view that the spatiotemporal world is (absolutely) *directioned* in the sense that, for any two nonoverlapping spatiotemporal entities, one exists "in a certain (absolute) direction" of the other; and the latter exists "in the opposite (absolute) direction" of the former. If our world is a *spatiotemporal* world (and not a world that is, as Leibniz and Newton believed, both spatial *and* temporal), then it is false that *spatial* directional relations exist. And there would exist no *temporal* relations such as *exists [or occurs] simultaneously with, came [or tenselessly comes] into existence before, will occur [or tenselessly occurs] after*. Instead, in a spatiotemporal world, persisting four-dimensional spatiotemporal entities would be tenselessly and eternally extended spatiotemporally. And, each entity would be "in a certain (absolute) spatiotemporal direction" from other entities. That is, in a spatiotemporal world, *spatiotemporal* directional relations would obtain among nonoverlapping spatiotemporal entities . . . regardless of whether there exist regions of spacetime that these spatiotemporal entities occupy.

In their textbook on relativity, physicists Edwin F. Taylor and John Archibald Wheeler presuppose Spatiotemporal Substantivalism, but they appear to presuppose as well that there would be specific (absolute) directions in a spacetime world—that there would exist spatiotemporal directional relations that obtain among the objects that pepper spacetime. Leading up to their explanation of the calculation of the dimensions of a region of spacetime, Taylor and Wheeler use a thought experiment involving the release of ball bearings at each end of a train car:

> "Region of spacetime." What is the precise meaning of this term? The long-narrow railway coach in the example served as a means to probe spacetime for a limited stretch of time and in one or another single direction in space. It can be oriented north–south, or east–west, or up–down. Whatever the orientation, the relative acceleration of the tiny ball bearings released at the two ends can be measured. For all three directions—and for all intermediate directions—let it be found by calculation that the relative drift of the two test particles is half the minimum detectable amount or less.[17]

Taylor and Wheeler make the point that the relative acceleration of the ball bearings is measurable regardless of the *directional* orientation of the train car—regardless of the *directional* relation that one end of the car bears to the other end.

As I argued in chapter 3, neither *Spatial* Directionalism nor a directionalist account of dimensionality would imply that there is a "correct (or privileged) outside perspective" with respect to whether a specific absolute directional relation corresponds with "absolute North" or "absolute South," or with "absolute up" or "absolute down," or with "absolute left" or "absolute right." By the same reasoning, the existence of *spatiotemporal* directional relations would not imply that a certain relation (or set of relations) corresponds with spacetime's "absolute North" or "absolute up" or "absolute left": An object in a four-dimensional *spacetime* world would involve at least two of each of four sets of *spatiotemporal* directional relations, but this would *not* imply that one of these four sets corresponds with "absolute North" or "absolute West" or "absolute up" or "absolute left."

Informally, Spatiotemporal Directionalism is the view that our world is *spatiotemporally directioned*—that there exist spatiotemporal directional relations that obtain among whatever spatiotemporal entities there may be:

D7.1 x is a spatiotemporal entity =Df It is logically possible that there exists something to which x bears a spatiotemporal directional relation.

D7.2 Spatiotemporal directional relation *sdn* obtains between x and y =Df Where *sdn* is a spatiotemporal directional relation, either (i) x bears *sdn* to y or (ii) y bears *sdn* to x.[18]

Spatiotemporal entities—for example, persisting electrons, persisting humans, persisting stars—would be the bearers of spatiotemporal directional relations, and such a relation would obtain between two such entities when one bears that relation to the other. And, if persisting spatiotemporal entities are perduring entities, then at least one spatiotemporal directional relation would obtain between any two nonoverlapping spatiotemporal parts of any given persisting spatiotemporal entity.

One can formulate Spacetime Directionalism precisely by invoking *spatiotemporal* directional relations (instead of *spatial* directional relations) and by recasting several of chapter 3's "spatial" analyses as "spatiotemporal" analyses:[19]

D7.3 p is a possible location for spatiotemporal entity x =Df p is a set of all and only those things that are spatiotemporal directional relations such that, for any member of p, *sdn*, it is logically possible both (i) that x bears *sdn* to something y and also (ii) that x bears any other member of p to something other than y.

D7.4 p is the actual spatiotemporal location of x =Df p is a possible location for x; and x and p are such that, for any y and for any spatiotemporal directional relation *sdn*, if x bears *sdn* to y, then *sdn* is a member of p.

D7.5 Spatiotemporal entity x involves spatiotemporal directional relation *sdn* =Df There is a set p that is the actual spatiotemporal location of x, and *sdn* is a member of p.

Informally, a *possible location* for some spatiotemporal entity is the set of all and only those spatiotemporal directional relations such that the entity *could* bear any two of those relations to other spatiotemporal entities. And, a spatiotemporal entity's *actual location* is a possible location that includes as members any spatiotemporal directional relation that the entity bears to other entities. To say of a spatiotemporal entity that it *involves* a certain spatiotemporal directional relation is to say that that relation is a member of the set that *is* that entity's actual location.

In terms of involvement, *Spatiotemporal Directionalism* is the view that all spatiotemporal entities involve at least one spatiotemporal directional relation:

> STD The world is *spatiotemporally directed*: There exist spatiotemporal directional relations; and any spatiotemporal entity that exists involves at least one spatiotemporal directional relation.

If the favored interpretation of Relativity is correct and if there exists a persisting materially solid sphere, then this four-dimensional sphere would be composed of a (four-dimensional) succession of nonoverlapping three-dimensional, spherical spatiotemporal parts. And, if STD is also correct, then any one of the four-dimensional sphere's three-dimensional, spherical spatiotemporal parts would involve the spatiotemporal directional relations that the spatiotemporal part's top hemisphere bears to its bottom hemisphere, that its right hemisphere bears to its left hemisphere, and that its front hemisphere bears to its rear hemisphere. (The spherical spatiotemporal part would also involve, of course, the opposite relations as well.) And, the three-dimensional sphere that is a spatiotemporal part of the (persisting) four-dimensional sphere would also involve all of those indefinitely many spatiotemporal directional relations (and their opposite relations) that obtain between it and other nonoverlapping spatiotemporal parts of the spatiotemporally extended sphere. Finally, the spatiotemporally extended sphere would itself involve all the spatiotemporal directional relations (and their opposite relations) that it bears to other nonoverlapping spatiotemporally extended (four-dimensional) objects *and* to the spatiotemporal parts of *those* persisting objects.

As Graham Nerlich has noted, merely countenancing spatiotemporal relations does itself leave open the question of whether regions of spacetime exist:

> Everything real is *related*, spatio-temporally, to every other real thing (i.e. bodies). This deletes reference to places and times. But does it delete space and time themselves? To answer that question is to take up a perspective on the debate between space-time substantivalists and their main opponents, relationists.[20]

If spatiotemporal relations *do* obtain among spatiotemporal entities, then it would remain an open question as to whether there also exist regions of spacetime occupied by those entities such that those same spatiotemporal relations would also obtain among those regions of spacetime. Thus, in the same way that (as explained in chapter 2) *Spatial* Directionalism is consistent with both Spatial Substantivalism *and* Spatial Relationalism, Spatiotemporal Directionalism is similarly consistent with both Spatiotemporal Substantivalism *and* Spatiotemporal Relationalism: Claiming that the world is (absolutely) directioned leaves open the possibility that there exist regions of substantivalist spacetime among which spatiotemporal directional relations obtain.

Spacetime Directionalism and Spatiotemporal Substantivalism. Given that the favored interpretation of Relativity involves Spatiotemporal Substantivalism, defenders of the favored interpretation who are also sympathetic with STD should allow that regions of spacetime can themselves be the bearers of (absolute) spatiotemporal directional relations. Consider again a person who is born in New York and dies eighty years later in Seattle. One of the person's wrinkled, hunched, infirm spatiotemporal parts in Seattle would (tenselessly) occupy a certain region of spacetime; and both that old-person spatiotemporal part *and* the region of spacetime that it occupies would (tenselessly) bear a certain set of spatiotemporal directional relations to one of the person's chubby, healthy, baby spatiotemporal parts in New York *and* also to the region of spacetime that that baby spatiotemporal part (tenselessly and eternally) occupies. Similarly, that baby spatiotemporal part *and* the region of spacetime that it (tenselessly and eternally) occupies would bear the opposite spatiotemporal directional relations to both the old-person spatiotemporal part *and* to the region of spacetime that it (tenselessly and eternally) occupies.

DIRECTIONALIST SPATIOTEMPORAL RELATIONALISM

A committed relationalist who is persuaded that the world is both spatiotemporal (rather than spatial *and* temporal) *and* directioned may invoke spatiotemporal directional relations to formulate a directionalist/relationalist theory of spacetime. Such a theory that would allow one to skirt ontological commitment to substantivalist spacetime by reformulating *spacetime*-implying claims as *non*-spacetime-implying claims cast in terms of spatiotemporal entities and the *spatiotemporal* directional relations that they exhibit. And, such a theory would allow the relationalist to preserve certain substantivalist intuitions; for example, that it is possible that all spatiotemporal entities are moving uniformly and that the inverted, mirror-image counterpart of the actual spatiotemporal world *could* have existed instead. *Directionalist Spatiotemporal Relationalism* is the result of combining Spatiotemporal Relationalism with Spatiotemporal Directionalism:

DSR   It is false that *spacetime* or regions of *spacetime* exist; and spatiotemporal directional relations do exist such that reasonable claims that imply that *spacetime* or regions of *spacetime* exist can be reformulated as claims (involving spatiotemporal entities and spatiotemporal directional relations) that clearly do not imply that *spacetime* or regions of *spacetime* exist.

One would expect DSR to enjoy advantages analogous with those that the Directionalist Theory of *Space* enjoys:

- DSR would allow one to explicate spatiotemporal dimensionality in terms of spatiotemporal directional relations.
- DSR would be consistent with the view that our world is a world of more than four spatiotemporal dimensions—a view endorsed by physicists who endorse string theory.[21]
- Without invoking substantivalist spacetime, one could use spatiotemporal directional relations to reformulate the revised version of Conjoining—a noncircular, commonsense answer to the Special Composition Question.[22]
- Without invoking substantivalist spacetime, one could use spatiotemporal directional relations to reformulate the reductivist theories of boundaries and holes that were (in chapter 6) formulated in terms of spatial directional relations.
- Most importantly perhaps, DSR is ontologically simpler than Spatiotemporal Substantivalism: DSR does *not* imply the existence of substantivalist *spacetime*.

Though the considerations above favor DSR, the question remains: Are there other overriding considerations that favor instead the existence of substantivalist *spacetime*? If our world *is* a spatiotemporal world, *is* it possible to reduce all reasonable *spacetime*-implying claims to non-*spacetime*-implying claims involving spatiotemporal entities and the spatiotemporal directional relations? More to the point: *Are* there findings of modern physics that provide a compelling case for the favored interpretation of Relativity and thereby for the existence of substantivalist spacetime?

THE CASE FOR SPATIOTEMPORAL SUBSTANTIVALISM

Consider five defenses of Spatiotemporal Substantivalism—five reasons to conclude that what is known about the world *cannot* be adequately explained without commitment to substantivalist *spacetime*. The Directionalist

Spatiotemporal Relationalist has a ready objection to the first of these four defenses, but the other four defenses are more resistant to relationalist replies.

Positing substantivalist spacetime can explain the deformation of the surface of the water in Newton's spinning bucket. Physicist Brian Greene notes that Einstein abandoned his hope of developing a relationalist theory of spacetime because, in part, the relationalist cannot allow that the surface of the water in a spinning bucket would be concave in a universe that contained no spatiotemporal entities other than the spinning bucket of water and its parts:

> Mach advocated a relational conception of space: for him, space . . . was not itself an independent entity. Initially, Einstein was an enthusiastic champion of Mach's perspective. . . . But as Einstein's understanding of general relativity deepened, he realized that it did not incorporate Mach's ideas fully. According to general relativity, the water in Newton's bucket, spinning in an otherwise empty universe, would take on a concave shape, and this conflicts with Mach's purely relational perspective, since it implies an absolute notion of acceleration.[23]

This defense of Spatiotemporal Substantivalism is the spatiotemporal version of the fourth defense of *Spatial* Substantivalism addressed in chapter 4: The best explanation for the deformation of the water's surface would be that the spinning bucket of water undergoes *absolute* motion—that, as it spins, the bucket of water undergoes actual changes of spacetime locations while undergoing *no* change of spatiotemporal relations that it bears to other spatiotemporal entities (given that no other such entities exist).[24]

Hartry Field underscores the force of Newton's "bucket" argument, claiming that "[w]hat Newton's bucket argument shows most directly is not that we must adhere to a substantivalist view of space-time, but that we need a notion of absolute acceleration." And, Field claims, the difficulty in formulating a concept of absolute acceleration is simply a special case of the more general "problem of quantities"—the problem of specifying *relationally* the positions of spatial entities at a given time. Field builds a case that, at best, it is problematic whether a relationalist can provide an adequate account of acceleration *or* adequate accounts of the relation *is twice the distance from* or *is twice the acceleration of* or *is orthogonal to the acceleration of*.[25]

Appealing to rotation in particular, John Earman argues that the relationalist cannot deliver the requisite concept of absolute motion, claiming that "[r]elativity theory, in either its special or general form, is more inimical to a relational conception of motion than is classical physics" and that "it is difficult to find a space-time structure that is recognizably relativistic and that fails to make the existence of rotation an absolute."[26] Maudlin observes, however, that "[i]nertial forces do not disbar relationalism, but they do place requirements on the nature of the relations posited";

and he claims that "[r]elations that are sufficiently rich can provide the means of explaining inertial effects."[27] Consistent with Maudlin's claim that a Spatiotemporal Relationalist needs "sufficiently rich" relations to preserve absolute motion, there is a concept of absolute motion available to the relationalist who is willing to embrace spatiotemporal *directional* relations.

As explained in chapter 4, the relationalist who accepts *Spatial Directionalism* can, without invoking substantivalist *space*, formulate an account of absolute motion cast in terms of (absolute) *spatial* directional relations: "A spatial entity undergoes absolute motion when it occupies successively different *spatial* locations—when it exhibits successively different *spatial* directional relations throughout a temporal interval." Similarly, without invoking substantivalist *spacetime*, the *Spatiotemporal* Directionalist can offer an account of absolute motion cast in terms of (absolute) spatiotemporal directional relations. Such a Spatiotemporal Directionalist would claim that each part of a rotating bucket occupies successively different *spatiotemporal* locations—each bucket part exhibits successively different *spatiotemporal* directional relations. For example, the spatiotemporal directional relations that the bucket's left half would bear to its right half are other than those that the left half would bear to its right half after the bucket spins a quarter turn. In short, then, absolute motion involves a spatiotemporal entity's changing its spatiotemporal orientation: It involves a successive change of the spatiotemporal directional relations that obtain among a spatiotemporal entity's parts as the entity moves.[28] And, this sort of successive change of spatiotemporal directional relations would occur even in a possible world in which the bucket of water and its parts are the only spatiotemporal entities that exist.[29]

If plausible interpretations of Special and General Relativity *do* require the concept of absolute motion, is the Spatiotemporal Relationalist's directionalist account of absolute motion adequate for the task? I am not myself competent to answer this question and leave it to physicists and philosophers of physics to determine whether there *is* a plausible interpretation of Relativity that incorporates a directionalist/relationalist account of absolute motion to provide acceptable accounts of acceleration, rotation, and particle positions. Consider now four defenses of substantivalist spacetime that cannot be dismissed easily by invoking spatiotemporal directional relations.

Positing substantivalist spacetime allows for a plausible reductivist account of fields. If the most promising interpretation of Relativity involves explaining the behavior of various spatial entities in terms of *fields* (e.g., the electromagnetic field), then there would be a significant reason to posit substantivalist spacetime: One could reduce *field*-implying claims to claims about the causal properties exemplified by (or causal predicates applied to) spacetime points or regions of spacetime points. Earman and Field both address this matter:

> In the nineteenth century the electromagnetic field was construed as the state of a material medium, the luminiferous ether; in postrelativity theory it seems that the electromagnetic field, and indeed all physics fields, must be construed as states of [space-time manifold] M.[30]

> [T]he natural way for a substantivalist to view a field is not as some giant physical object which occupies all (or virtually all) of space-time, but as not an entity at all. . . . [A] field theory is simply a theory that assigns causal properties to space-time regions directly. . . To put it in a less ontologically inflationary way, it is simply a theory that employs causal predicates that apply directly to space-time regions. This means that acceptance of a field theory is not acceptance of any extra ontology beyond space-time and ordinary matter.
>
> . . .
>
> For instance, in electromagnetic field theory we assign to each point in space-time an electromagnetic intensity, irrespective of whether this point is occupied by matter.[31]

In short, then, if the best physics does require embracing *the electromagnetic field*, then ontological economy can be achieved by affirming that the field *is* (substantivalist) spacetime points and the electromagnetic values that the points exhibit.[32] Consider three replies available to the Spatiotemporal Relationalist who resists this "field" defense of substantivalist *spacetime*: The Spatiotemporal Relationalist could (a) reject field theory altogether, (b) endorse a nonreductivist field theory, or (c) offer a reductivist theory of fields formulated in terms of spatiotemporal relations and spatiotemporal entities, *not* in terms of substantivalist spacetime.

Hartry Field addresses the question of whether rejecting field theory is plausible:

> [I]t seems to me that for a physical theory to accord with anything reasonably called relationalism, that physical theory can not be a field theory. Instead of predicting and explaining the behavior of matter in terms of fields, i.e., properties of (unoccupied as well as occupied) regions of space-time, a relationalist physical theory would have to predict and explain the behavior of matter in terms only of that matter and other matter (e.g., the matter that a substantivalist might intuitively think of as "generating" the relevant aspects of the field). A physical theory which is relationalist in this sense is called an action-at-a-distance theory.[33]

As Field notes, to resist the view that one should posit substantivalist spacetime in order to preserve field theory, the Spatiotemporal Relationalist could simply deny the existence of fields and then replace field theory with some theory of physics that could explain the behavior of a given material entity X in terms of *other* spatiotemporal entities. The relationalist explanation, then, could *not* involve appealing to, say, the values of the electromagnetic field *where* (or *near where*) X is located; rather, the explanation would be cast in terms spatiotemporal entities other than X—spatiotemporal entities are at a distance from X and not even "indirectly connected" with X by way of an intermediate field that lies between them and X. And, such at-a-distance material entities would be, as Field puts it, "'generating' the relevant aspects of the field."

Is there any hope of replacing field theory with a plausible *field*-less theory of physics? A critic of *field*-less theories may object that action-at-a-distance theories are problematic because they are mysterious: They imply that a particular material entity X's behavior is explicable not by local field values but, mysteriously, by material entities that are at a distance from X—entities that somehow, at a distance, "generate" or "exhibit" the "relevant aspects" of a field needed to explain X's behavior. *Are* action-at-a-distance *field*-less theories hopelessly mysterious? Though he expresses concern regarding the prospects for success, Field does himself leave open the question of whether a plausible *field*-less theory can be developed:

> It seems to me that the question of whether it is possible to find an interesting way to dispense with fields is quite a fascinating one. The difficulties in doing this are quite considerable, but I certainly would not want to dismiss out of hand the possibility that it can be done. To my mind, the commitment to doing it is perhaps the most interesting consequence of the relationalist position.[34]

I lack the expertise to address intelligently the question of whether one can replace field theory with a relationalist action-at-a-distance theory. So, I leave open the question of whether the Spatiotemporal Relationalist can resist the "field" defense of substantivalist *spacetime* by rejecting the existence of fields.

A second reply available to the Spatiotemporal Relationalist committed to resisting the "field" defense of substantivalist spacetime is to agree with the field theorist that fields exist and then offer a nonreductivist theory of fields, claiming that fields are not identical with regions of (substantivalist) spacetime that exhibit field values. Rather, on this nonreductivist view, a field would itself be "a giant object . . . with a part corresponding to each space-time region"[35] and the behavior of a given spatiotemporal entity would be explained by field values exhibited by the field itself, not by field values exhibited by regions of substantivalist spacetime nor by other at-a-distance spatiotemporal entities that "generate" the "relevant aspects" of a field.

The obvious concern regarding this second reply is that the Spatiotemporal Relationalist who endorses nonreductivist fields thereby sacrifices ontological economy in order to skirt ontological commitment to substantivalist spacetime: Unlike a Spatiotemporal Relationalist committed to spatiotemporal entities and relations alone, the relationalist who posits spatiotemporal entities, spatiotemporal relations, *and* fields lacks an ontological advantage over the Spatiotemporal Substantivalist who posits spatiotemporal entities, spatiotemporal relations, *and* spacetime (especially if the commitment to spacetime affords the physical theory additional explanatory power). As Field puts the point, the Spatiotemporal Relationalist who endorses nonreductivist fields "'saves' relationalism only by trivializing it: on an interesting version of relationalism, fields are just as much a fiction as is space-time."[36] Maudlin observes that "insofar as the [substantivalist/relationist] debate is motivated by the considerations voiced by Leibniz and Newton, the ... debate has at last resolved itself into a purely verbal dispute." Maudlin notes that, according to the substantivalist who reduces fields to regions of spacetime, "the world is at base a manifold of spacetime points which support fields, in which the fields inhere." And, according to the Spatiotemporal Relationalist who posits a nonreductivist field (a *plenum*), "the world is a field or set of fields (and perhaps particles) which instantiate spatiotemporal relations, in which the relations inhere." Maudlin concludes that "[t]his difference of expression cannot, in the contexts of [the General Theory of Relativity], be promoted into any further dispute about the physical or metaphysical facts."[37]

Regarding the "field" defense of substantivalist spacetime, there is a third reply available to the Spatiotemporal Relationalist who is committed to field theory, to ontological economy, *and* to Spatiotemporal Directionalism: Such a Spatiotemporal Relationalist could offer a reductivist theory of fields in terms of spatiotemporal entities and spatiotemporal relations alone, reducing *field*-implying claims to claims about the field values linked with particular directionalist/relationalist possible locations.[38] (Per D7.3, a possible location for a given spatiotemporal entity would be a set of all and only those spatiotemporal directional relations such that, for any two of those relations, it is possible that the entity simultaneously bears those relations to two other spatiotemporal entities.) With respect to the electromagnetic field, for example, the Spatiotemporal Relationalist could suggest something like the following: "There is electromagnetic intensity N and a possible location P such that, if there exists a spatiotemporal entity X such that P *is* or is 'near' X's actual spatiotemporal location [per D7.4], then X would be subject to electromagnetic intensity N."[39]

Is such a directionalist/relationalist reductivist field theory plausible? The critic may begin by reminding such a reductivist that *field*-less action-at-a-distance theories are objectionable because they *are* mysterious—because they imply that a particular material entity's behavior is explicable by material entities *at a distance*, not by field values exhibited by something *local*. Then, the critic may charge that the directionalist/relationalist's reductivist field theory is no less mysterious: Though such a field theory would preserve the attribution of field values to local unoccupied *possible locations*, there would be, strictly speaking, no *local* entity of any sort that could be the bearer of those values. Thus, the directionalist/relationalist's reductivist field theory would be saddled with the same mystery as an action-at-a-distance theory—the mystery of explaining a particular material entity's behavior not by local field values but by at-a-distance material entities that would be "'generating' the relevant aspects of the field."

The tempting reply to this objection is that the directionalist/relationalist/reductivist field theory is no more mysterious than the *substantivalist*/reductivist theory:

> The substantivalist who reduces *field*-implying claims to claims about the field values of spacetime points theory is committed to the mysterious view that *immaterial* spacetime points can *be* the bearers of field values and that these *immaterial* points-with-field-values *can* causally affect the behavior of *material* spatiotemporal entities. If anything, invoking a causal connection between spacetime points and material entities is more objectionable that invoking action-at-a-distance explanations involving material entities only. (I consider additional concerns about spacetime points later in this chapter.)

Is the substantivalist's reductivist field theory so mysterious that one should reject it? Or, is the action-at-a-distance mystery even greater—so great that it warrants rejecting both a *field*-less theory of physics *and* the directionalist/relationalist's reductivist field theory? Again, I lack the expertise to speculate intelligently about how one should answer these questions. I leave to physicists and philosophers of physics the task of determining whether field theory *is* an essential part of any plausible theory of physics and whether, ultimately, field theory *requires* substantivalist spacetime instead of directionalist/relationalist possible locations. And by leaving *these* questions to others, I thereby leave to others the question of whether field theory does or does not give one a reason to side with Spatiotemporal Substantivalism instead of Spatiotemporal Relationalism.

Positing substantivalist spacetime allows for a reductivist account of gravity. If one does *not* reify spacetime, then one may be tempted to countenance the existence of at least one type of *force*: Gravity. So conceived, gravity would *not* itself be a physical entity, but it *would* be *something*—a force—that "resides" in physical entities with mass.[40] And the degree to which the gravity that resides in one physical entity would "exert pull" on a second physical entity would be a function of the first entity's mass and the distance between that entity and the second. The gravity that "resides" in an object of great mass would exert far more "pull" on nearby objects than it does on distant objects; and the gravity that "resides" in an object of great mass "exerts" twice the "pull" on other objects as the gravity that "resides" in an object of half its mass.

Consider two puzzling questions that motivate the search for an alternative to Newton's gravity-is-a-force view. (a) Gravity is neither a physical object nor an abstract entity; it is a *force*. But what, exactly, *is* a *force*? (b) And, what, exactly, is the mechanism by which a force would "push" or "pull" on another object? How, *exactly*, would the gravity that "resides" in the Sun "reach" across more than 90 million miles to "pull" Earth toward the Sun? Would the Sun emit "rays" of gravity that strike distant objects and then "reel them in" toward the Sun? Or, would gravity function more like a vacuum cleaner, "sucking" distant objects toward the object in which the gravity "resides?" Physicist Greene summarizes these concerns about the Newtonian gravity-is-a-force view:

> For well over two centuries, Newton's universal law of gravity had done an impressive job at predicting the motion of everything from planets to comets. Even so, there was a puzzle that Newton himself articulated: How does gravity exert its influence? How does the Sun influence the Earth across some 93 million miles of essentially empty space? Newton had provided an owner's manual allowing the mathematically adept to calculate the effect of gravity, but he was unable to throw open the hood and reveal how gravity does what it does.[41]

The mysteries that surround the gravity-*is*-a-force view should motivate one to take seriously the reductivist account of gravity that Einstein's General Relativity affords: Talk of gravity is reducible to talk about substantivalist *spacetime*. In their textbook on spacetime physics, physicists Edwin Taylor and John Wheeler summarize Einstein's view:

> Galileo and Newton viewed motion as properly described with respect to a rigid Euclidean reference frame that extends through all space and endures for all time. . . . Within this ideal space of Galileo and Newton there acts a mysterious force of gravity, an interloper from the world of physics, a foreign influence, not described by geometry. In contrast, Einstein says that there is no mysterious "gravitation," only the structure of spacetime itself.[42]

> . . .

> Why all this mention of "curvature of spacetime?" [Because]—in the case of gravitation—one has a simple interpretation for observations in terms of the geometry of curved spacetime. One no longer has to assume that the world is built of spacetime plus some mysterious magical "physical" force of gravitation, foreign to and added to spacetime.[43]

Simply put, Einstein claimed that, by endorsing the existence of *malleable* substantivalist spacetime, one can explain why certain spatiotemporal entities "are attracted to" other spatiotemporal entities without committing oneself to the existence of Newton's mysterious gravitational force. Physicist Greene offers a sketch of how one can explain "gravitational attraction" in terms of malleable spacetime:

> Einstein showed that much as a warped wooden floor can nudge a rolling marble, space and time can themselves warp, and nudge terrestrial and heavenly bodies to follow the trajectories long ascribed to the influence of gravity.[44]

> It's as if matter and energy imprint a network of chutes and valleys along which objects are guided by the invisible hand of the spacetime fabric. That, according to Einstein, is how gravity exerts its influence.[45]

Physicist Wheeler summarizes the view more directly: "Spacetime tells matter how to move; matter tells spacetime how to curve."[46] Imagine dropping a cue ball onto a relatively soft wooden floor and then dropping a heavy bowling ball a half-meter away. Although the cue ball would dent the floor where it lands and would also slightly deform the floor surrounding the dent, the bowling ball would create a much deeper dent and would deform the surrounding floor so markedly that the cue ball would roll toward the bowling ball, following the warp of the floor. According to General Relativity, gravitational attraction involves something similar: It involves a physical object's deforming the region of spacetime that it occupies and also deforming the surrounding spacetime region such that smaller entities would then follow the curvature of the deformed spacetime, falling toward the larger object.[47] To say that the Sun exerts gravity on Earth is to say that the massive Sun deforms spacetime, which causes Earth to fall toward the Sun, following the curvature of the Sun-deformed spacetime. The existence of (malleable) substantivalist spacetime is, then, essential to the account of gravity that the favored interpretation of General Relativity affords. Consider corroborating evidence for this reductivist theory of gravity.

According to Einstein's spacetime theory of gravity, the massive Sun would greatly distort the region of spacetime that surrounds it. So, if this theory is correct, one would expect to find evidence that light waves bend as they travel near the Sun (or near other massive entities)—they would bend as they follow the curvature of the distorted spacetime. As Stephen Hawking reports in his popular writing, the first observational evidence of matter bending starlight was reported a century ago:

> [General Relativity] was confirmed in a spectacular fashion in 1919 when a British expedition to West Africa observed a slight bending of light from a star

passing near the sun during an eclipse. Here was direct evidence that space and time are warped.

Einstein's general theory of relativity transformed space and time from a passive background in which events take place to active participants in the dynamics of the universe.[48]

There is also evidence that the gravitational bending of light has created illusions for observers on Earth. As Eddington observed, the Sun can deflect another star's light, causing it to appear that the other star—the source of the light—was located in a slightly different position. Scientists have also reported that the bending of a galaxy's light has created the illusion of there being *two* or more galaxies.[49]

Summary. There is adequate observational evidence that the deflection of starlight occurs and that large spatiotemporal masses (e.g., the Sun) play a causal role in this phenomenon. The view that gravity is a classical force that resides in objects of mass cannot explain the measured amount of deflection of light, and this is a mark against Newton's gravity-is-a-force-in-*non*-malleable-space view. One *can* explain, however, the measured deflection of light by claiming that substantivalist spacetime exists, that objects of mass warp spacetime (in proportion to the masses of the objects), and that light thereby follows the measured curved spacetime path.

Positing substantivalist spacetime better explains the expansion of the universe. Astrophysicists tell us that there is adequate evidence that the universe is rapidly expanding in the sense that galaxies and their constituents are moving ever faster and further away from one another.[50] But what is the best explanation for this expansion? Jim Al-Khalili summarizes a prevailing answer to this question:

> [A] common misconception of the expansion of the Universe is that all the other galaxies are flying through space away from our own. This is wrong. In fact, it is the empty space in between the galaxies that is stretching.[51]

Brian Greene has offered a more detailed popular account:

> In 1927, Lemaître applied Einstein's equations of general relativity not to objects within the universe, like stars and black holes, but to the entire universe itself. The result knocked Lemaître back on his heels. The math showed that the universe could not be static: The fabric of space was either stretching or contracting, which meant that the universe was either growing in size or shrinking.
>
> When Lemaître alerted Einstein to what he'd found, Einstein scoffed. . . . So certain was Einstein that the universe, as a whole, was eternal and unchanging,

> that he not only dismissed mathematical analyses that attested to the contrary, he inserted a modest amendment into his equations . . . to ensure that the math would accommodate his prejudice.
>
> And prejudice it was. In 1929, the astronomical observations of Edwin Hubble . . . revealed that distant galaxies are all rushing away. The universe is expanding. Einstein gave himself a euphemistic slap in the forehead, a reprimand for not trusting results coming out of his own equations.
>
> . . .
>
> The seemingly unavoidable conclusion is that the universe we see emerged from a phenomenally tiny speck that erupted, sending space swelling outward—what we now call the Big Bang.[52]

Astrophysicists have evidence, then, that galaxies and their constituents are *not* expanding through spacetime in the same way that pieces of shrapnel are propelled through the air by a grenade's explosion. Rather, the evidence suggests that galaxies and their constituents are expanding away from one another because each spatiotemporal entity occupies a specific region of spacetime, each specific spacetime region is a part of the whole of spacetime, and the whole of spacetime is itself expanding. As the whole of spacetime stretches and expands, each constituent region and its occupant, if any, expands away from other constituent regions and *their* occupants.

Consider a version of a standard example used to characterize two conceptions of how galaxies and their constituents could move apart as the universe evolves from the Big Bang. If there were two ants asleep on the surface of a semi-inflated round balloon, there would be two methods for causing the ants to move away from one another. First, one could gently nudge the ants, provoking them to stir and crawl away from one another. Second, without waking the ants, one could gently add air to the balloon: As the balloon inflates, the sleeping ants will nonetheless move away from one another as the balloon's surface does itself expand.[53] Astrophysicists would tell us that the second type of ant movement corresponds with the means by which galaxies expand.

Summary. Astrophysicists have adequate evidence that the universe is expanding and that the best explanation for this expansion is that (*expandable*) substantivalist spacetime exists: Spatiotemporal entities are expanding away from one another because they occupy regions of substantivalist spacetime that are themselves moving away from one another as the whole of substantivalist spacetime expands.

The detection of gravitational waves is evidence that substantivalist spacetime exists. The previous two sections were explications of two strong abductive arguments for substantivalist spacetime: By postulating malleable, expandable substantivalist spacetime, one can best explain what gravity is and its "pull" across great distances; and one can best explain how it is that galaxies, their constituents, and any other spatiotemporal entities are expanding. More recently, there has emerged evidence for substantivalist spacetime that purports to be more direct: The detection of "gravitational waves"—that is, the detection of rippling substantivalist spacetime itself.

A decade before physicists claimed to have detected a gravitational wave, physicist Greene explained the nature of such waves and how one might detect them:

> General relativity predicts that, just as a trampoline assumes a fixed, warped shape if you stand perfectly still, but heaves when you jump up and down, space can assume a fixed, warped shape if matter is perfectly still . . . , but ripples undulate through its fabric when matter moves to and fro. Einstein came to this realization . . . when he used the newly fashioned equations of general relativity to show that . . . matter racing this way and that (as in a supernova explosion) produces gravitational waves. And since gravity is curvature, a gravitational wave is a wave of curvature. Just as tossing a pebble into a pond generates outward-spreading water ripples, gyrating matter generates outward-spreading spatial ripples; according to general relativity, a distant supernova explosion is like a cosmic pebble that's been tossed into a spacetime pond. . . . [U]nlike . . . water waves that travel *through* space[,] . . . gravitational waves travel *within* space. They are traveling distortions in the geometry of space itself.[54]

According to the favored interpretation of Relativity, substantivalist spacetime is not only the thing *where* spatiotemporal entities are located, but it is a thing that can be warped, that can be propelled outward (moving spatiotemporal entities with it), *and* that can ripple when disturbed. Ripples of spacetime would emanate outward from a disturbance; and the bigger the disturbance, the stronger the waves that would ripple across spacetime. This substantivalist view also implies that the rippling of spacetime causes the rippling of whatever spatiotemporal entities occupy rippling regions of spacetime. Thus, if a wave of spacetime causes Earth to ripple and if one detects Earth's ripple, then one would thereby be detecting the spacetime wave that caused Earth's ripple. The detection of a spacetime wave *is* the detection of a gravitational wave: "[Since] according to general relativity, curved spacetime is gravity, a wave of curved spacetime is a wave of gravity,"[55] that is, a gravitational wave.

Greene explains how one would devise a device to detect a gravitational wave.[56] Like sound waves and the various electromagnetic waves, gravitational waves would have frequencies, and the frequency of gravitational waves would be "the number of peaks and troughs per second" that travel across spacetime. Just as different musical instruments generate sound waves with different frequencies, different astrophysical phenomena would generate gravitational waves with different frequencies, and this is the key to detecting gravitational waves. Scientists have been able to calculate, for example, the specific pattern of frequencies of gravitational waves that a supernova would produce and know that this pattern is different from the calculated frequencies of the gravitational waves that the collision of two orbiting black holes would produce. Thus, to construct a device that can determine whether Earth ripples with one of these telltale patterns of frequencies is to construct a device that measures the pattern of the gravitational waves that caused Earth's rippling. Such a device, then, would provide evidence both that there exists rippling substantivalist spacetime *and* that there did occur eons ago a specific type of astrophysical event that caused the telltale spacetime rippling detected by the device.

Scientists have constructed two such devices to measure Earth's rippling—two Laser Interferometer Gravitational-Wave Observatories [LIGOs]. Though these two large devices—one in Louisiana and the other in Washington—became operational in 2002, they detected nothing and were shut down in 2010 for a five-year upgrading to enhance their sensitivity. Two days after their reactivation in 2015, within mere milliseconds of one another, each LIGO detected the telltale patterns of frequencies attributed to gravitational waves generated more than a billion years ago by the collision of two black holes.

Summary. Physicist Greene does himself note the problem that the report of the detection of gravitational waves—that is, the detection of ripples of substantivalist spacetime—poses for a relationalist account of spacetime:

> If general relativity fully incorporated Mach's [relationalist] ideas, then the "geometry of space" would merely be a convenient language for expressing the location and motion of one massive object with respect to another. Empty space, in this way of thinking, would be an empty concept, so how could it be sensible to speak of empty space wiggling? Many physicists tried to prove that the supposed waves in space amounted to a misinterpretation of the mathematics of general relativity. But in due course, the theoretical analyses converged on the correct conclusion: gravitational waves are real, and space *can* ripple.[57]

There is compelling scientific evidence that each LIGO detected the Earth's rippling; and the preferred interpretation of Relativity implies that the rippling was caused by the rippling of the spacetime that Earth occupies. There is evidence, then, that substantivalist spacetime *is* a thing that *is* capable of propagating waves that can cause Earth's rippling; because relationalists reject the existence of substantivalist spacetime, the relationalists lack an obvious explanation for Earth's rippling.

Summary of the case for substantivalist spacetime. Setting aside the "absolute motion" and "field theory" defenses, there are at least three significant scientific reasons to countenance the existence of substantivalist spacetime: Postulating the existence of spacetime affords one a plausible theory of gravity, an explanation of the expansion of the universe, and an explanation of why Earth ripples. Oliver Pooley summarizes the case for Spatiotemporal Substantivalism:

> [S]ubstantivalism is recommended by a rather straightforward realist interpretation of our best physics. This physics presupposes geometrical structure that it is natural to interpret as primitive and as physically instantiated in an entity ontologically independent of matter.[58]

Consider now reasons to be skeptical about the existence of substantivalist spacetime.

REASONS TO BE SKEPTICAL ABOUT SPATIOTEMPORAL SUBSTANTIVALISM

As Graham Nerlich notes, "Space-time is metaphysically peculiar, perhaps even bizarre (immaterial yet with concrete relations to concrete things and, most worrying, elusive to perception)."[59] What, exactly, would lead one to conclude that substantivalist spacetime *is* "peculiar" or "bizarre?"

Is it logically possible that a location *can warp, expand, or ripple?* To expose one troubling implication of endorsing substantivalist spacetime, consider again the nature of Newton's concept of "absolute space." According to the Newtonian account of spatial location, the location for any physical object or other spatial entity is a region of substantivalist *space*—an entity composed of a dense array of "immovable" spatial points.[60] Space, on this view, would neither be a physical object nor any other spatial entity; rather, it would be a categorically different thing that would be the "cosmic container" for physical objects and any other entities that have location—the fixed, unchanging thing that would be *where* located entities are located.

Suppose that one bends the top half of a vertical 8-cm wire at a 90-degree angle. According to the Newtonian concept of substantivalist space, it is logically possible that one bends the top half of the wire, but it would be logically *im*possible that anything could move the region of space *where* the top half of the wire *was* located before it was bent. When the wire's top half bends, its former location—a region of spatial points—may itself become unoccupied, but it would otherwise remain unchanged and would continue to bear exactly the same spatial relations to the region of spatial points occupied by the wire's bottom half. Bending space, then, would be as impossible as folding a monad in half.

According to the standard conception of space as the thing *where* located entities are located, if a massive physical object moves into an unoccupied region of space, it would be logically impossible for that object to deform that region or to compress the spatial points that constitute that region. Detonating an atomic bomb within a region of space would blast debris outward across space, but the explosion could neither possibly disrupt the ordering of spatial points where the explosion occurred, nor could it blast the surrounding space outward from the explosion. To claim that a region of space could be propelled outward to occupy a *place* that it did not occupy earlier would be absurd: On the Newtonian view, regions of space would *be* places; they are not things that could *occupy* places or *change* places. Similarly, though an explosion can produce soundwaves (i.e., ripples of air; the successive compression and rarefaction of air molecules), the explosion could not possibly cause ripples of *space*—the successive compression and stretching of spatial points as they successively occupy different *locations*.

The above survey of implications of the Newtonian view that spatial locations are regions of substantivalist space exposes the bizarre nature of the Einsteinian view that locations are regions of substantivalist spacetime—locations that would be essentially bendable, expandable, and capable of rippling. One may be tempted to argue, then, that it is logically impossible that substantivalist spacetime exists:[61]

> To claim that a region of spacetime is bendable, expandable, or capable of rippling is to claim that a region of spacetime points can be bent or propelled or caused to rise to a *location* that it did not occupy earlier. If the Sun distorts nearby spacetime, it would cause a certain region of spacetime to bend into a location that it did not occupy earlier. If the Big Bang propelled galaxies *and* substantivalist spacetime outward, they would be moving into *locations* that nothing—not even spacetime—occupied earlier. If the collision of black holes causes the successive squeezing and stretching of regions of spacetime, it causes spacetime points to move into locations that they did not occupy earlier and will later cease to occupy. Thus, this substantivalist spacetime view is absurd: If regions of spacetime points exist, they would *be* locations that could not themselves bend, expand, or ripple, moving *from* one location *to* another. Thus, Spatiotemporal Spacetime should be rejected.

Leaving open the question of whether this objection is adequate, consider a second metaphysical concern about substantivalist spacetime.

Is it logically possible that regions of spacetime and physical objects causally interact? As physicists Jeremy I. Pfeffer and Schlomo Nir note, the standard interpretation of relativity implies that there is an intimate causal relationship between spacetime and physical objects:

> A body's presence affects the shape of the spacetime in its neighbourhood and that of its geodesics. This, in turn, affects the trajectories of other bodies moving through the spacetime.[62]

More specifically, referring back to the evidence for substantivalist spacetime, the favored interpretation of Relativity implies that physical objects distort spacetime, and the distorted spacetime can in turn affect the movement of other objects, causing them to deviate from the paths they would have otherwise followed. The favored interpretations also imply that the Big Bang caused spacetime to expand and that this ongoing expansion of spacetime is in turn moving galaxies apart much like a moving river moves an innertube. And, according to the favored interpretations of SRL and GRL, a supernova and other astrophysical phenomena can cause spacetime itself to ripple, and when the ripples of spacetime (i.e., gravitational waves) reach a spacetime region occupied by a planet, the spacetime ripples cause the planet to ripple as well.

The favored interpretation of Relativity treats the causal interaction between substantivalist spacetime and physical objects as unproblematic. But is it? To develop the metaphysical concern regarding this alleged causal connection, consider again Newton's view that substantivalist *space* and physical objects are categorically different entities that cannot interact causally. On the Newtonian view, space is conceived as the inert, nonphysical entity that *is where* physical objects are located and *where* all causal interaction involving physical objects takes place. If a lightning bolt's striking a dead tree ignites a forest fire, the discharge of electrons, the tree, and the burning forest would all be located *in* space, not *in* a physical object; and the regions of space that the discharged electrons, tree, and forest occupy could not possibly play a causal role in the lightning strike's causing the fire. Claiming that a region of space causally interacts with a physical object that *occupies* that region is on a par with claiming that a swarm of monads causally interacts with the square root of 11.

Similarly, then, if it is at least dubious that *space* and physical objects can causally interact, then it is at least dubious that *spacetime* and physical objects can causally interact when one conceives of substantivalist spacetime as a nonphysical *something* that is *where* physical objects or object-involving events are located. A metaphysician may be tempted to argue, then, that it is logically impossible that substantivalist spacetime and physical objects interact causally:

> Most metaphysicians are skeptical that a physical body could causally interact with a Cartesian soul given that, unlike any *physical* entity, a soul would be a *non*-physical entity without mass, charge, spin, or any type of material composition. If they exist, spacetime points and regions of spacetime points would also be entities without mass, charge, spin, or any type of material composition; so, one should be similarly skeptical that a physical object could warp a region of spacetime. If one is skeptical that a surgeon's scalpel could halve a Cartesian soul, then one should be similarly skeptical that the Sun can deform a region of spacetime, displacing that region's points.
>
> By the same token, one should also be skeptical that a warped region of spacetime could causally affect the moon's orbit or the path taken by a star's light. At an amusement park, when a child allows his kiddy car to drift, the track's curbs keep the car on the track: The wheels of the car bump the low steel curbs that border the winding track, and the curbs "push back" on the wheels. Equal and opposite reaction, these forces must be the same. There is, then, a plausible explanation of how the curvy steel curb that borders the curvy track causally affects the car's path: The curb and go-cart are physical objects, and their surfaces make contact such that the curb "applies a direction-changing force to the wheel." Curved spacetime, however, is not the sort of thing that can make contact with a planet's moon or ray of light to "transfer force." So how, *exactly*, could substantivalist spacetime—a non-physical, massless, surface-less entity—cause a deviation in a moon's orbit or the bending of starlight?
>
> The favored interpretation of Relativity implies that spacetime is expandable, but how can this be? Bread dough expands as it rises, increasing in volume. The larger volume, however, is not due to an increase in the amount of dough: Dough molecules do *not* come into being *ex nihilo*! Rather, the existent dough molecules are pushed further apart by the gas produced by the fermenting yeast. Keeping this in mind, how is it logically possible that spacetime expands? To allow spacetime to increase in volume to fill "the empty space" where no spacetime yet exists, is there a constant coming into being *ex nihilo* of indefinitely many spacetime points? Or, does spacetime expand in the sense that the spacetime points are pushed further apart? But how could *this* happen? Would "gaps" appear between spacetime points? (But what would such a gap be? An empty region of *space*? A *place* where no space exists? A region of *meta-spacetime* where not even *spacetime* exists?)
>
> Finally, the favored interpretation of Relativity implies that when a region of spacetime ripples, it causes any physical object that occupies (i.e. overlaps with)

that region to ripple as well. But how *can* the rippling of a nonphysical *something* cause the rippling of an overlapping physical object. One who is skeptical that a Cartesian soul could causally interact with one's physical body should be no less skeptical if told that one's soul overlaps with one's brain. And, how, *exactly*, is it possible that spacetime *ripples*? One can explain coherently what the rippling of water involves: Water ripples when certain water molecules are repeatedly pushed upward and then allowed to fall downward—when certain water molecules are pushed upward from one (substantivalist or relationalist) location to a higher (substantivalist or relationalist) location and are then allowed to return to their original location below. A particular region of spacetime cannot, in *this* sense, ripple: A particular region of spacetime *is* a location, so it would be absurd to insist that it is pushed from one location to a higher location and then allowed to return to its lower location. In short, the concept of a location's changing its location is incoherent.

Leaving open the question of whether this second metaphysical objection is adequate, consider a third reason to be skeptical that substantivalist spacetime exists.

Is there a coherent difference between curved, expanded, or rippling spacetime and spacetime that is entirely undistorted? The above explication of two potential metaphysical objections to substantivalist spacetime glosses over an implication of the favored interpretation of Relativity. The favored interpretation implies that, *strictly*, there is *no* region of spacetime R such that R is *not* bent, expanded, or rippled and then *becomes* bent, expanded, or rippled. That is, the favored interpretation of Relativity implies that (a) *no* straight, vertical region of spacetime points can be warped by a physical object such that those *same* points constitute an L-shaped region, (b) *no* compact region of spacetime points can expand such that those *same* points compose a "much larger" region of spacetime, and (c) *no* set of particular spacetime points can undulate such that those same points repeatedly rise upward and then fall downward. The reason that the favored interpretation of Relativity implies (a), (b), and (c) is that the favored interpretation involves a perdurantist theory of persistence: Because it is false that spacetime *and* (substantivalist) times both exist, it is false that any specific spacetime point exists *at multiple times*; rather, the favored interpretation implies that regions of spacetime points are the only spatial/temporal locations that exist and that all spacetime points exist tenselessly and eternally—spacetime points are all equally real. Thus, it is impossible that a certain set of spacetime points is configured in a particular way and is *later* configured in a different way; whatever the configuration of a certain set of spacetime points, that is the configuration of that set of points tenselessly and eternally.

The only sense in which there can exist a persisting spacetime point is the sense in which there can exist a four-dimensional spacetime point composed of spatiotemporal *parts* each of which is a (nonpersisting) spacetime point. And the only sense in which there can exist a persisting region of spacetime is the sense in which there can exist a four-dimensional spacetime region composed of spatiotemporal *parts* each of which is a three-dimensional spacetime region composed of different (nonpersisting) spacetime points. If two nonoverlapping three-dimensional regions of spacetime points are spatiotemporal parts of a persisting region of spacetime, then no spacetime point that belongs to one of the three-dimensional regions belongs to the other.

With the perdurantist concept of persistence, then, the favored interpretation of Relativity does allow for a sense in which a region of spacetime can bend, expand, or ripple. For example, to claim that straight, rod-shaped (i.e., |-shaped) region of spacetime bends into a curved C-shaped region is to claim that there is persisting (i.e., four-dimensional) region of spacetime points such that (i) one of its spatiotemporal parts is a rod-shaped three-dimensional region of spacetime points and (ii) a later spatiotemporal part—a spatiotemporal part that bears such-and-such spatiotemporal directional relation to the rod-shaped part—is a C-shaped three-dimensional region composed of altogether *different* spatiotemporal points.

Similarly, according to the favored interpretation of Relativity, the sense in which spacetime can expand is the sense in which there is a persisting (i.e., four-dimensional) region of spacetime points such that (i) one of its earlier spatiotemporal parts is a relatively compact subregion composed of specific spacetime points and (ii) one of its later spatiotemporal parts is a much larger subregion composed of *different* spacetime points. Finally, the favored interpretation allows that spacetime undulates in the sense in which there is a persisting (i.e., four-dimensional) region of spacetime composed of spatiotemporal parts that are alternatingly smaller/flatter and larger/bowed where each subregion is composed of an entirely different set of spacetime points. Keeping in mind the perdurantist accounts of what the warping, expanding, and rippling of substantivalist spacetime involves, consider a third metaphysical concern about substantivalist spacetime.

Imagine two *spacetime manifolds*, M and M*, which are both arrays of four-dimensionally extended, "equally real" spacetime points.[63] Assume that M is the actual world: M would be an expanding array—each spatiotemporal part would be larger than any earlier part and smaller than any later part—and various spatiotemporal entities would occupy regions of M and causally contribute to there existing (four-dimensional) regions of warped spacetime and (four-dimensional) regions of undulating spacetime that M includes. Assume that M* is a *non*expanding spacetime manifold occupied by *no* spatiotemporal entities capable of distorting spacetime such that M would be a densely packed array of spacetime points that includes *no* bent or undulating regions of spacetime. Now, compare spacetime manifolds M and M*, disregarding any spatiotemporal entities that either manifold contains.

Though M is an expanding array—"smaller on one end, larger on the other"—and M* is not, what other difference could there possibly be between M and M*? Would not M and M* be otherwise indistinguishable from one another? One may be tempted by this answer: "Unlike M*, M includes regions of warped and undulating spacetime!" But how would a manifold with warped and undulating subregions differ from a manifold without such subregions? Would not the arrangement of spacetime points in M and M* be identical? Would not the expanding and nonexpanding manifolds be otherwise structurally equivalent? There would seem to be no sense in which warped or undulating regions of spacetime points could be arranged differently in M than the nonbent, nonundulating regions in M*. Imagine any three-dimensional slice of spacetime points across undistorted (four-dimensional) M*: The slice would be a dense array of spacetime points such that between any two there is another. Now compare this three-dimensional slice of M* with any three-dimensional slice of (four-dimensional) actual world M: Exactly like the slice of undistorted M*, the slice of M would also be a dense array of spacetime points such that between any two there is another. It would be impossible that, in a warped or rippled three-dimensional slice of M, there would exist *more* spacetime points than in the slice of M*; and it would be impossible that the spacetime points in a warped or rippled slice of M would be more "*densely* packed" or "bunched together" or "more compact" than the spacetime points in the undistorted slice of M*.

To develop this point differently, imagine that, in order to explain gravity and gravitational waves in the classroom, an astrophysics professor decides to construct models of both M and M*: One model of a universe with gravity and gravitational waves, and a second model of a universe without. The professor's plan is to construct these two models by filling half-meter cube-shaped glass containers with (dimensionless) mereological simples. The professor first constructs the model of M*, which will represent a spacetime manifold without distortions: She pours indefinitely many simples into the glass cube to cover its bottom with a first layer of simples one-simple thick; then, she adds layer after layer of indefinitely many simples until the glass cube is filled with a "perfectly even," cube-shaped array of simples such that, between any two, there exists a third. The result is a dense array of simples *without* any distorted or rippled subregion. A problem arises when the professor prepares for the construction of the second model: How *does* one construct a model of M that would both represent a spacetime manifold *with* distorted and rippled regions *and* that would differ structurally from the completed model of M*? It would seem that any dense arrangement of simples in the glass cube that represents M would be identical to the dense arrangement of simples in the glass cube that represents M*! After all, suppose that the professor attempts to construct the model of M by adding to the glass cube layers of indefinitely many simples that exhibit an obvious depression (representing the significant depression of spacetime that a star would cause). Then, the professor adds wavy layers of simples to represent gravitational waves. In the end, when she finishes filling the container with indefinitely many densely packed mereological simples, the model of M would be structurally indistinguishable from M*: There would be no depressions or waves of simples in the model of M; each glass cube would contain indefinitely many densely packed mereological simples such that, between any two, there exists a third. Any two-dimensional half-meter square slice of either model would be structurally indistinguishable from any two-dimensional half-meter square slice of the other: Each slice would contain indefinitely many simples in a perfectly dense array. Appealing to this thought experiment, a metaphysician may be drawn to the following argument: "If there can be no structural difference between the model of M* and the model of M, then there can be no structural difference between undistorted spacetime and spacetime that is warped or rippled. And, if there can be no structural difference between an undistorted spacetime manifold and a manifold that includes distortions and waves, then the concept of distorted spacetime is dubious, if not incoherent, which poses a serious problem for the favored interpretation of General Relativity."

ARE THE METAPHYSICIAN'S OBJECTIONS COMPELLING?

Do the three metaphysical objections developed in the preceding section constitute a compelling case against the favored interpretation of Relativity? If so, can the case against the favored interpretation be parlayed into a defense of Directionalist Spatiotemporal Relationalism? As Newton wrote (when at a loss to explain how the force of gravity could instantaneously affect objects at a great distance), "[This question] I have left to the consideration of my readers."[64] Though I ultimately leave open the question of whether the three objections constitute a compelling *a priori* case against malleable substantivalist spacetime (and thereby leave open the question of whether the case against the favored interpretation of Relativity can be used to defend DSR over Spacetime Substantivalism), a metaphysician more confident than I may press the following Metaphysical Defense of Directionalist Spatiotemporal Relationalism:

MDR 1. The favored interpretation of Relativity is correct only if it is possible that substantivalist spacetime has the capacity to warp, expand, or ripple.
2. There are compelling *a priori* reasons to conclude that it is logically impossible that substantivalist spacetime has the capacity to warp, expand, or ripple.
3. Therefore, there are compelling reasons to conclude that the favored interpretation of Relativity is not correct. (from 1,2)
4. If there are compelling reasons to conclude that the favored interpretation of Relativity is not correct, then there is no positive reason to endorse Spacetime Substantivalism instead of Directionalist Spacetime Relationalism.
5. Therefore, there is no positive reason to endorse STS instead of DSR. (from 3,4)
6. DSR is ontologically simpler than STS.
7. If there is no positive reason to endorse STS instead of DSR and if DSR is ontologically simpler than STS, then it is reasonable to believe that DSR is correct.
8. Therefore, it is reasonable to believe that DSR is correct. (from 5,6,7)

The Spacetime Substantivalist who accepts the metaphysical objections to malleable spacetime could resist this defense of relationalism by objecting to MDR-4, rejecting malleable spacetime and thereby rejecting the favored interpretation of Relativity. And, by rejecting the favored interpretation of Relativity, the substantivalist would no longer be able to defend substantivalist spacetime abductively by claiming that the existence of (malleable) spacetime would best explain the nature of gravity, the expansion of the universe, or the alleged detection of gravitational waves. So, defending STS by rejecting MDR-4 would involve quite a cost for the substantivalist. Without the abductive arguments for (malleable) spacetime that an appeal to the favored interpretation of Relativity affords, the Spatiotemporal Substantivalist incurs the burden of building a compelling case for nonmalleable substantivalist *spacetime*, and the burden would be significant: Without the appeal to the favored interpretation of Relativity, the substantivalist must now argue both that the case for (nonmalleable) substantivalist spacetime is stronger than the case for Spatiotemporal Relationalism *and* that there is adequate evidence for siding with the existence of substantivalist *spacetime* instead of substantivalist *space*. After all, if the substantivalist abandons the case for a (malleable) spacetime manifold composed of "static" spacetime points, why not embrace instead an *enduring* three-dimensional array of substantivalist spatial points—an array of spatial points that do themselves exist at every *time*. Rejecting MDR-4, then, could well take the substantivalist back to the arguments addressed earlier in chapter 4.

There is a second cost that the Spatiotemporal Substantivalist incurs by rejecting MDR-4, and it is a cost also incurred by the relationalist who defends DSR by appealing to MDR: By endorsing MDR-1 and MDR-2, one is thereby committed to *rejecting* the favored interpretation of Relativity—a significant cost indeed! Given the explanatory and predictive power of the favored interpretation of Relativity, I would myself be reluctant to defend DSR by siding with *a priori* metaphysical objections to malleable substantivalist spacetime *unless* I could site a replacement interpretation of Relativity that is at least as plausible and powerful as the favored interpretation. Whether a viable alternative interpretation is available is at best problematic, but this may not deter those metaphysicians who place more confidence in the *a priori* objections to malleable spacetime than I do. Such metaphysicians, emboldened perhaps by Ned Markosian's reply to the Special Relativity Objection to Presentism, may defend MDR and thereby DSR even though they *lack* a viable replacement interpretation of Relativity. To develop this defense of MDR, consider first a sketch of the Special Relativity Objection to Presentism and Markosian's bold reply.

Roughly, the Special Relativity Objection is this:[65]

> Presentism—the view that anything that exists *presently* exists—implies that there is a unique present such that, for *any* two objects that (presently) exist anywhere in the universe, there is a present *time* at which those two objects exist. According to Special Relativity, however, there is *no* unique present: A given time can be present relative to one frame of reference but not relative to another, and there can be no unique "correct" frame of reference relative to which a given time is or is not *the* present time. Thus, the compelling evidence on behalf of Special Relativity is compelling evidence *against* a unique "correct" frame of reference, and this in turn is compelling evidence *against* Presentism's implication that there *is* a privileged time that *is* the present.

Markosian's reply to the Special Relativity Objection involves distinguishing between a "philosophically loaded" interpretation of Special Relativity and a "philosophically lean" interpretation:[66]

> SRL^+ = an interpretation of SRL that includes philosophical presuppositions rich enough to entail that there can be no unique present.
> SRL^- = an interpretation of SRL that is empirically equivalent to SRL^+, but which is so philosophically lean that it fails to entail that there can be no unique present.

Markosian's reply to the SRL objection to Presentism is this:

> Although I agree that there seems to be a great deal of empirical evidence supporting [SRL⁺], I think it is notable that the same empirical evidence supports [SRL⁻] equally well. And since I believe there is good *a priori* evidence favoring [SRL⁻] over [SRL⁺], I conclude that [SRL⁻] is true and that [SRL⁺] is false.[67]

Markosian, then, courageously rejects the favored interpretation of SRL (i.e., SRL⁺): By distinguishing SRL⁻ and SRL⁺, he can enthusiastically endorse the empirically confirmed mathematics that underlies SRL⁻ while appealing to *a priori* metaphysical arguments (on behalf of a unique present) to conclude that physicists should abandon SRL⁺ as the correct interpretation of the mathematics.

The Directionalist Spatiotemporal Relationalist may be tempted by a similar strategy to strengthen the Metaphysical Defense of DSR:

> The empirical evidence on behalf of the mathematics that underlies SRL and GRL is compelling. But how one should *interpret* the mathematics is another matter altogether. The favored interpretations of SRL and GRL—that is, the favored interpretation of Relativity—do imply that malleable and expandable substantivalist spacetime exists; but there are compelling *a priori* metaphysical objections to such spacetime. Given these objections, then, one should side with ontologically leaner Directionalist Spatiotemporal Relationalism and conclude that physicists should reject the "philosophically rich" favored interpretation of Relativity, which is philosophically loaded on behalf of Spatiotemporal Substantivalism.

The Directionalist Spatiotemporal Relationalist should be wary of this strategy for the same reason that Presentists should be wary of Markosian's stand against the favored interpretation of SRL. Though I would myself be reluctant to stake the rejection of the favored interpretation of SRL solely on an *a priori* metaphysical defense of a unique present, I would be more sympathetic with Markosian's stand against SRL⁺ if it were coupled with a replacement interpretation of SRL—a physical theory that rivals SRL⁺ with respect to explanatory and predictive power and that has implications less objectionable than SRL⁺'s (alleged) implication that there exists *no* unique present.[68] Similarly, then, I would not myself rest the rejection of the favored interpretation of Special and General Relativity solely on *a priori* metaphysical objections to malleable and expandable substantivalist spacetime! After all, the favored interpretation has served physicists and astronomers extraordinarily well with respect to explanatory and predictive power. I would, however, be less reluctant to press the case against the favored interpretation of Relativity if I had in hand at least the core of a plausible replacement interpretation—the core of a physical theory that would be empirically equivalent with the favored interpretation of Relativity (i.e., a theory that would share the favored interpretation's robust explanatory and predictive power). And, I would want the replacement interpretation well enough developed to make clear that *its* implications are no more bizarre than the view that substantivalist spacetime is malleable and expandable.[69]

Are there any promising empirically equivalent alternatives to the favored interpretation of Special and General Relativity—alternatives that are no more bizarre than the favored interpretation? The answer to this question rests well beyond the scope of this book, but a review of the SRL/GRL defense of malleable and expandable substantivalist spacetime suggests that there are at least three serious problems to be addressed when entertaining a replacement for the favored interpretation of Relativity—a replacement that is entirely *spacetime*-less or that implies that spacetime exists but is nonmalleable and nonexpandable.

The first problem involves gravity. The favored interpretation of GRL offers a theory of gravity that can explain, for example, the bending of light as it passes by the Sun: "Gravity *is* deformed substantivalist spacetime; so, the Sun's gravity bends light in the sense that the Sun deforms spacetime, and light follows the deformed spacetime's curvature." *How*, exactly, could an adequate, empirically equivalent replacement interpretation of the mathematics of GRL explain this bending of light without an appeal to substantivalist spacetime? To answer this question, should one revisit the view that gravity *is* a force that "resides" in physical objects and that "reaches out" to "exert pull instantaneously" on neighboring objects? This would be unfortunate: Einstein found his concerns regarding the gravity-is-a-force-at-a-distance view so serious that they provided him with a motivation for developing GRL with its gravity-is-warped-spacetime implication. If, however, one resists ontological commitment both to bizarre malleable substantivalist spacetime and to a mysterious gravitational *force*, then what *plausible* full-blown empirically equivalent replacement interpretation of Relativity *could* explain the apparent bending of light?[70]

The second problem involves the manner in which the universe appears to be expanding. The favored interpretation of Relativity provides an explanation of how this expansion happens: "The Big Bang caused substantivalist spacetime itself to expand. And given that spatiotemporal entities (including galaxies and their constituents) are located in spacetime, these entities are thereby moving apart as well, 'carried along' by the expanding spacetime itself." If one insists on an empirically equivalent replacement interpretation of Relativity that does not require ontological commitment to expandable substantivalist spacetime, then how *could* the alternatives plausibly explain the accelerating expansion of galaxies and their constituents? Should one consider replacing ontological commitment to substantivalist spacetime with ontological commitment to a "propelling force" that pushes the galaxies apart? But what would a "propelling force" *be*? How would such a force come into being and become "activated" to propel physical entities further apart? And how, *exactly*, could a force that is not itself a physical entity causally affect both massive and microscopic physical entities across vast distances? In the end, would the existence of a "propelling force" be any less problematic than the existence of malleable and expandable spacetime? And just how likely is it that the existence of such a force could *be* an implication of a plausible, *comprehensive*, empirically equivalent replacement interpretation of the philosophically lean mathematics that underlies SRL and GRL?

A third problem for a replacement interpretation of Relativity involves the evidence that the Earth itself undergoes rippling. With an appeal to gravitational waves, the favored interpretation of Relativity provides an explanation of the detected rippling of Earth: "Gravity *is* substantivalist spacetime. So, when an astrophysical event disturbs spacetime, it causes spacetime ripples (i.e. *gravitational* waves), and these spacetime ripples cause the rippling of those physical objects that occupy those regions of rippling spacetime." If one insists on a replacement interpretation of Relativity that is free of ontological commitment to malleable substantivalist spacetime that can undulate, then how, *exactly*, can one *plausibly* explain the detected Earth ripples in terms of a comprehensive physical theory that is empirically equivalent with the favored interpretation of Relativity? Should one consider seriously the view that a cataclysmic astrophysical event (e.g., a supernova explosion) could "exude" an "*explosion*" force" that "travels" for eons before it "makes contact" with Earth, causing Earth ripples? But what would such an "explosion force" *be*? And how would a cataclysmic event involving a physical object bring such a *force* into being? How, *exactly*, for a billion years, would a force "travel" across vast distances? And in what sense, *exactly*, would such a *force* "make contact" with the physical Earth, thereby causing the rippling of a massive object? If, one does manage to construct plausible answers to these questions about the workings of an "explosion force," *would* one also be able to incorporate the existence of an "explosion force" within a physical theory that is comprehensive enough to be empirically equivalent with the favored interpretation of Relativity?

Summary. The fate of Directionalist Spatiotemporal Relationalism or any other version of Spatial Relationalism rests in large part with the answer to this question: Is there an empirically equivalent replacement for the favored interpretation of Relativity that does not involve ontological commitment to *malleable substantivalist spacetime*—a comprehensive physical theory that is consistent with the mathematics that underlies Relativity and that can explain gravity, the expansion of the universe, and Earth's rippling without positing the existence of *malleable substantivalist spacetime* that has the capacity to undulate? The answer to this question rests well beyond the scope of this book and is, ultimately, a question to be addressed by physicists and philosophers of physics.[71]

GOING NOWHERE: THE PROSPECTS

Tim Maudlin is less than enthusiastic about the effort to resolve the dispute between relationalists and substantivalists:

> [W]e could, with Leibniz and Mach, somehow try to eliminate space-time as an entity altogether, but it is unclear what either the motivation or the prospects of such a project are.[72]

Contrary's to Maudlin's assessment, there *is* motivation for defending Directionalist Spatiotemporal Relationalism or some other relationalist theory of location. The relationalist theory could well be an ontologically simpler theory of *location* given that it skirts ontological commitment to substantivalist spacetime. And, there is ample philosophical motivation to avoid ontological commitment to substantivalist spacetime given that regions of such spacetime would themselves be metaphysically bizarre: They would be nonphysical *locations* that can change shape, that can expand, that can undulate, and that can interact causally with physical objects and events. What *are* the prospects regarding Directionalist Spatiotemporal Relationalism or some other relationalist theory of location?

What are the implications for a relationalist theory of location if one discovers a spacetime-*less replacement for the favored interpretation of Relativity?* If physicists and philosophers of physics someday endorse a plausible, comprehensive, *spacetime*-less interpretation of Relativity, then this would strengthen the case for a relationalist theory of location (e.g., the Directionalist Theory of *Space* or Directionalist *Spatiotemporal* Relationalism): If gravity, the expansion of the universe, and Earth's rippling *can* be explained without invoking substantivalist spacetime that is warped, expandable, or undulating, then apparently there would be no compelling reason to believe that substantivalist spacetime exists. And, if one can invoke spatial or spatiotemporal directional relations to reduce, respectively, reasonable *space-* and *spacetime*-implying claims to directionalist claims that clearly do not imply the existence of substantivalist space or spacetime, then one could reasonably defend a relationalist theory of location with an appeal to ontological simplicity and a reminder that distortable substantivalist spacetime would be utterly bizarre.

What are the implications for a relationalist theory of location without *a spacetime-less replacement for the favored interpretation of Relativity?* If physicists and philosophers of physics continue to assure us that the case for distortable substantivalist spacetime is compelling, then consider why the relationalist appeal to spatiotemporal directional relations will not have been in vain.

The evidence for the favored interpretation of Relativity is evidence that physical entities or events *do* have metaphysically bizarre substantivalist spacetime locations that can *overlap* ("spatiotemporally coincide with") physical entities and that can be bent, that can expand, and that can undulate. Ontological commitment to *malleable spacetime* would be more akin to ontological commitment to luminiferous aether than to Newton's absolute space.[73] And, one may then be tempted to infer that regions of substantivalist spacetime would *be* locations that can themselves *have* and *change* location as they bend, expand, or undulate. After all, the defender of distortable spacetime may insist that the following claims are accurate:

- The Sun and other physical objects warp regions of spacetime such that some of their subregions are crunched together and become located below or to the side of *where* they were located earlier.
- The Big Bang causes spacetime itself to expand outward from the Big Bang, expanding outward into places where nothing—not even spacetime—exists.
- Major astrophysical phenomena (e.g., the collision of black holes) send out ripples of spacetime such that, when a particular region undulates, it has a subregion that may repeatedly ripple—that is, that may repeatedly move upward to a place above its original location and then downward back to (or below) its original location.

Taken at face value, the above claims appear to imply that substantivalist spacetime would not be the only entity that *is* a location—that there would exist regions of *meta*-spacetime *where* every region of distortable spacetime is itself located. The defender of distortable spacetime may insist that the ontological implications of the claims above be made transparent:

- The Sun and other physical objects warp regions of spacetime: A physical object causes multiple spatiotemporal subregions to occupy regions of meta-spacetime that they did not occupy earlier; that is, a physical object would dislodge a particular subregion of spacetime, pushing it from one meta-spacetime region to a meta-spacetime region below or to the side of the meta-spacetime region *where* the spacetime region was located earlier.

- The Big Bang causes spacetime itself to expand outward from the Big Bang, expanding from one region of meta-spacetime to another and then into empty regions of meta-spacetime—regions of meta-spacetime where not even spacetime yet exists.
- Major astrophysical phenomena (e.g., collisions of black holes) send out ripples of spacetime such that, when a particular spacetime region undulates, it has a subregion that may repeatedly ripple—that is, that may repeatedly move upward to a meta-spacetime region that is above its original meta-spacetime location and then downward back to (or below) the meta-spacetime location that the spacetime region occupied initially.

One should be concerned, then, that commitment to malleable substantivalist spacetime may require commitment to substantivalist *meta*-spacetime as well! Fortunately, by dipping into the directionalist/relationalist's toolbox, the Spatiotemporal Substantivalist can sidestep such ontological inflation: By endorsing Spatiotemporal Directionalism, the Spatiotemporal Substantivalist can consistently use spatiotemporal directional relations—*not* substantivalist meta-spacetime—to explain the bending, expansion, and rippling of substantivalist spacetime. Here's how.

A directionalist/relationalist account of warped spacetime. According to the favored interpretation of Relativity, it is false that a particular region of spacetime is straight and then bent—it is false that a particular set of spacetime points constitutes a spacetime region that is straight at one time and then (composed of those *same* points) bent at a later time. After all, no spacetime point exists *at a time* given that, according to the favored interpretation, there are no *times*: A region of spacetime is *where* things are *spatiotemporally* located. Thus, to claim that a physical object bends a region of spacetime is to say, roughly, that one spatiotemporal part of that (four-dimensional) region is unbent while another spatiotemporal part is bent. In terms of spatiotemporal directional relations, the account of warped spacetime would be a bit more complicated:

> A four-dimensional region of spacetime R that "becomes bent" is composed, in part, of an unbent four-dimensional spatiotemporal subregion R_1, which is a spatiotemporal subregion that is composed of indefinitely many three-dimensional spatiotemporal parts such that there is a set S of, say, all and only those "east/west" (or "horizontal") spatiotemporal relations that obtain between any two of R_1's spatiotemporal parts. R "becomes bent" in the sense that R is also composed, in part, by another four-dimensional spatiotemporal subregion R_2, which is a spatiotemporal subregion that is composed of three-dimensional spatiotemporal parts, and it is false that the spatiotemporal directional relations that are

members of set S all obtain between any one of R_1's spatiotemporal parts and any one of R_2's spatiotemporal parts.

To make this clearer, refer to figure 7.1.

Let R represent a four-dimensional region of spacetime that "becomes bent." Where R_1 is a four-dimensional spacetime region that is a subregion of R, let R_{1a} and R_{1b} represent any two three-dimensional spatiotemporal parts—"spacetime slices"—of R_1 (and thereby of R); and let R_{2a} represent any three-dimensional spatiotemporal part of R_2 (and thereby of R) where R_2 is a four-dimensional subregion of R. Let sd_1 and sd_{13} represent any two "east/west" spatiotemporal directional relations that obtain between R_{1a} and R_{1b} and between any other two of the three-dimensional spatiotemporal parts of R_1. R "becomes bent" in the sense that it has both an "unbent" subregion and a "bent" subregion: Subregion R_1 is such that sd_1 and sd_{13} are among all of those "east/west" spatiotemporal directional relations that obtain between any two three-dimensional spatiotemporal parts of R_1; and R_2 is the "bent" subregion of R insofar as there are "east/west" spatiotemporal directional relations that obtain between any two spatiotemporal parts of R_1 (e.g., sd_1 and sd_{13}) that do not all obtain between any spatiotemporal part of R_2 (e.g., R_{2a}) and any spatiotemporal part of R_1. (For example, neither sd_1 nor sd_{13} obtains between R_{2a} and any spatiotemporal part of R_1.)

A directionalist/relationalist account of expanding substantivalist spacetime. If substantivalist spacetime *is* itself continuing to expand, then the Spatiotemporal Substantivalist can explain this ongoing expansion without invoking substantivalist meta-spacetime.

First, a reminder: According to the favored interpretation of Relativity, there is a sense in which the manifold of substantivalist spacetime could *not*

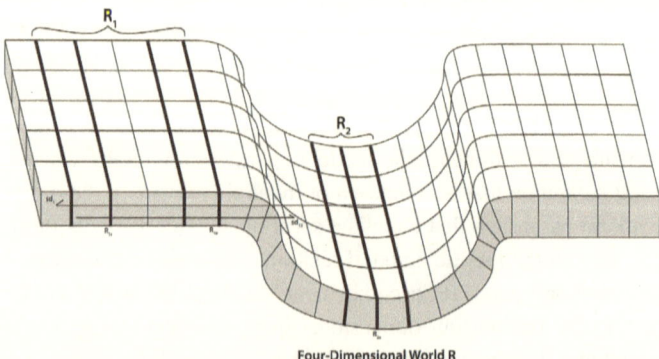

Figure 7.1 A "Bent" Region of Substantivalist Spacetime. *Source*: C. Frantom, J. Rich: White Roche LLC.

expand. Given that the favored interpretation implies a perdurantist theory of persistence, it would be false that the spacetime manifold composed of certain spacetime points is small *at one time* and then, composed of the *same* spacetime points, is larger *at a later time*. After all, according to the favored interpretations, regions of spacetime points *are* spatiotemporal locations, so claiming that anything, including spacetime points, exists *at a time* would be absurd. And, given that all spacetime points tenselessly and eternally exist, if any spacetime point x is after or before (or later or earlier than) spacetime point y, then x would not be identical with y. So, if a larger spacetime region is "after" a smaller spacetime region, the spacetime points that compose the former cannot be identical with the spacetime points that compose the latter.

According to the favored interpretation of Relativity, then, a persisting (four-dimensional) spacetime manifold expands in the sense that its earlier spatiotemporal parts would be smaller than its later spatiotemporal parts. More exactly, any given spatiotemporal part of an "expanding" (four-dimensional) spacetime manifold would be larger than any preceding spatiotemporal part and smaller than any spatiotemporal part that follows it. And, there would be causal connections between the (tenselessly and eternally occurring) Big Bang and each subsequent three-dimensional spatiotemporal part of the four-dimensional manifold. For example, the Big Bang would causally contribute to each subsequent three-dimensional manifold-slice's being larger than any manifold-slice that precedes it and smaller than any manifold-slice that follows it.

With this understanding of what an expanding universe would be like according to the favored interpretation of Relativity, one can appeal to spatiotemporal directional relations to explain the sense in which persisting (four-dimensional) spacetime "expands into regions where nothing exists." In the case of a four-dimensional world whose three-dimensional spatiotemporal parts are, following a Big Bang, "expanding" (i.e., are successively larger), there would be spatiotemporal directional relations that (a) larger three-dimensional spatiotemporal parts bear to certain earlier/smaller temporal parts and (b) those certain earlier/smaller spatiotemporal parts do *not* themselves bear to any other earlier/smaller spatiotemporal parts. For example, referring to figure 7.2, where EW represents a four-dimensional world in which substantivalist spacetime expands following the Big Bang, three-dimensional spatiotemporal part EW_{91} bears spatial directional relation sd_{897} to earlier three-dimensional spatiotemporal part EW_{72} (and to every spatiotemporal part between EW_{91} and EW_{72}; but there is no earlier/smaller spatiotemporal part to which EW_{72} bears sd_{897}. Thus, by appealing to spatiotemporal directional relations, the Spatiotemporal Substantivalist could skirt ontological commitment to regions of substantivalist meta-spacetime

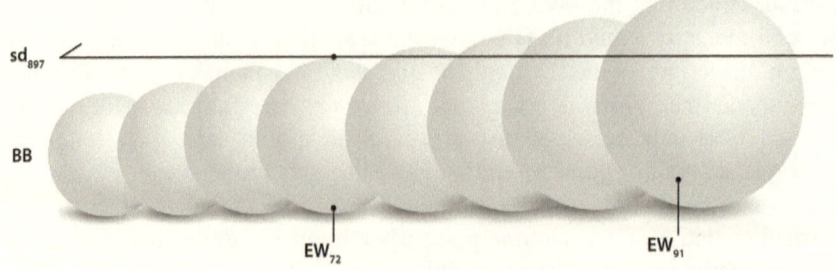

Figure 7.2 The Expansion of Substantivalist Spacetime. *Source*: C. Frantom, J. Rich: White Roche LLC.

while preserving the view that spacetime itself is (eternally) undergoing expansion.

A directionalist/relationalist account of undulating substantivalist spacetime. By appealing to spatiotemporal directional relations, the Spatiotemporal Substantivalist can explain gravitational waves—that is, the rippling of substantivalist spacetime—without claiming that a wave of spacetime is a spacetime region that, in some sense, moves back and forth between higher and lower regions of substantivalist meta-spacetime.

As figure 7.3 indicates, the substantivalist explanation of a spacetime ripple is similar to the substantivalist explanation of bent spacetime. According to the favored interpretation of Relativity, a spacetime ripple does *not* involve a particular region of spacetime that itself repeatedly moves upward and then downward with the passage of time. Rather, given the favored interpretation's implication that all spacetime points are "static" and "equally real," a ripple of spacetime—represented by G in figure 7.3—would be a "static" four-dimensional wavy-shaped region of spacetime with certain telltale patterns

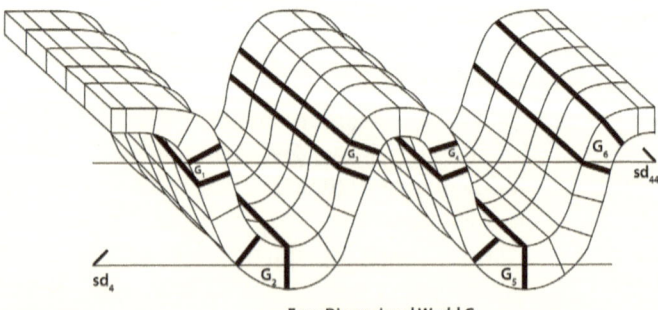

Figure 7.3 A Ripple in Substantivalist Spacetime. *Source*: C. Frantom, J. Rich: White Roche LLC.

of four-dimensional subregions. For example, in a rippled four-dimensional region of spacetime, there will be certain spatiotemporal directional relations that obtain between an earlier and a later four-dimensional subregion of the four-dimensional ripple even though those relations do *not* obtain between other four-dimensional subregions that are later than the former but earlier than the latter. With respect to figure 7.3, sd_4 obtains between three-dimensional subregions G_2 and G_5, but sd_4 does *not* obtain between G_3 and G_4 even though G_3 and G_4 are later than G_2 but earlier than G_5. And, spatiotemporal directional relation sd_{44} obtains between three-dimensional subregions G_1 and G_6 but not between G_2 and G_5 even though G_2 and G_5 are later than G_1 but earlier than G_6.

Conclusion. How far *should* one follow Leibniz down the road to *no*where? The honest metaphysician should be cautious, recognizing that the fate of a relationalist theory of location turns on future work by physicists and philosophers of physics. If their work results in replacing the currently favored interpretation of the mathematics that underlies Relativity with a new, comprehensive *spacetime*-less interpretation, then, by citing ontological simplicity and the bizarre nature of malleable substantivalist spacetime, then this present inquiry regarding Spatial Directionalism suggests that there would be hope that the directionalist/relationalist can forge a reasonable defense of the Directionalist Theory of Space.

If, however, physicists and philosophers of physics continue to argue that the most plausible interpretation of the findings of modern physics requires ontological commitment to malleable substantivalist spacetime, then the directionalist/relationalist would lack compelling cases for Directional Theory of Space, for Directionalist Spatiotemporal Relationalism, and for any other version of Spatial Relationalism. In this case, however, the directionalist/relationalist could find at least *some* consolation in pressing a defense of relationalism with respect to *meta*-spacetime, claiming that ontological commitment to malleable, expandable substantivalist spacetime does *not* also require ontological commitment to substantivalist meta-spacetime—that reasonable *meta-spacetime*-implying claims regarding *spacetime*'s distortion, expansion, or undulation can be reduced to *non*-meta-spacetime-implying claims involving nothing more than spatiotemporal entities, spacetime, and spatiotemporal directional relations.

NOTES

1. Friedman, *Foundations of Space-Time Theories*, p. 204.
2. In an essay that he dictated in 1917, Brentano cited the "strange doctrine" under which "time would be nothing other than a fourth dimension of space" such

that the spatial and the temporal would "no longer . . . be different concepts" but a "unified concept" that one might call "the spatio-temporal—of something having four dimensions." See Brentano, "What we can learn about space and time," p. 173–74.

3. STS allows that Maureen Donnelly and Antony Eagle would count as Spatiotemporal Substantivalists; see Maureen Donnelly, "Parthood and Multi-Location," *Oxford Studies in Metaphysics*, Vol. 5, ed. Dean Zimmerman (Oxford: Oxford University Press, 2010), p. 207: "I initially assume little more than that spacetime is a non-empty set of points, that regions are the non-empty subsets of spacetime." And, in the same collection, see Antony Eagle, "Location and Perdurance," p. 54: "One can safely think of regions [of spacetime] as being just whatever those things are that an object can be found in" and "regions are fundamentally made of points."

4. There are Spatiotemporal Substantivalists—the "Supersubstantivalists"—who side with reductivist accounts of objects cast in terms of regions of *spacetime*. See Nolan, "Balls and All," pp. 91–116. Nolan explores (but stops short of defending) Supersubstantivalism: "The primitive unit of spacetime is the region" (p. 92), and "we identify some of these regions with physical objects" (p. 95). Cf. David Lewis, *Parts of Classes* (Oxford and Cambridge, MA: Basil Blackwell, 1991), p. 75: "Suppose a material thing occupies a region of substantival space-time; it does not follow (though it just might be true) that the region is part of the thing, still less that the region and the thing are parts of each other and therefore identical." See also Sider, *Writing the Book of the World*, p. 292: Claiming that his "worldview's ontology contains only points of spacetime and sets," Sider accepts the implication that his ontology "contains no composite objects." For a succinct statement of reasons to endorse Supersubstantivalism, see Arntzenius, *Space, Time, and Stuff*, p. 182.

5. See also Arntzenius, *Space, Time, and Stuff*, p. 13: "According to Special Relativity, space and time are not separate entities—that instead there is one entity: spacetime." See also pp. 12–13.

6. Lawrence Sklar, *Space, Time, and Spacetime* (Berkeley, Los Angeles, London: University of California Press, 1977), p. 163.

7. Sider, *Four-Dimensionalism*, p. 43.

8. Greene, *The Fabric of the Cosmos*, pp. 59, 75; see also p. 58 for Greene's comparison of spacetime to a loaf of bread that can be sliced in different ways.

9. Cf. William R. Carter and H. Scott Hestevold, "On Passage and Persistence," *American Philosophical Quarterly* 31 (1994), 270. This formulation of Static Spacetime is the analogue of Carter and Hestevold's formulation of Static Time.

10. Theodore Sider and Alyssa Ney have argued that the favored interpretation of SRL implies that Transient Time is not correct. See Sider, *Four-Dimensionalism*, pp. 42–52; and see Ney, *Metaphysics: An Introduction*, pp. 143–44. At least one friend of Transient Time has argued instead that, although one should endorse the *uninterpreted* mathematics of SRL, one should *reject* its favored interpretation and insist on an (unspecified) interpretation that is consistent with Transient Time. See Ned Markosian, "A Defense of Presentism," *Oxford Studies in Metaphysics*, Vol. 1 (Oxford: Clarendon Press, an imprint of Oxford University Press, 2004), pp. 51, 76–78. Dean Zimmerman has boldly resisted the Relativity objections to Transient

Time; see Zimmerman, "The Privileged Present: Defending an 'A-theory' of Time," pp. 219–20. See also Zimmerman's "Presentism and the Space-Time Manifold," *The Oxford Handbook of Philosophy of Time*, ed. Craig Callender (Oxford: Oxford University Press, 2011), pp. 163–244. Nothing more will be said about the relationship between the favored interpretation of Relativity vis-à-vis temporal passage.

11. For a clear explication of Perdurantist and Endurantist (*and* Experdurantist) theories of persistence, see Sally Haslanger, "Persistence Through Time, Substantivalism," *The Oxford Handbook of Metaphysics*, pp. 317–20. See also Katherine Hawley, *How Things Persist* (Oxford: Clarendon Press, an imprint of Oxford University Press, 2001), pp. 9–20, 37–53.

12. Cf. Carter and Hestevold, "On Passage and Persistence," 276–78. Carter and Hestevold argue that Endurance is correct if and only if the Temporal Parity Thesis is false.

13. See Friedman, *Foundations of Space-Time Theories*, pp. 16–17. In describing Einstein's special theory of relativity, Friedman notes that it "falls short of the relationalist ambitions of Leibniz and Mach" given that (to preserve absolute acceleration and rotation) it invokes "four-dimensional inertial frames," which are "[f]rom the point of view of a thoroughgoing relationalism, . . . just as objectionable and 'metaphysical.'"

14. Kurt Vonnegut, *Slaughterhouse-Five* (1969), ch. 5, https://archive.org/details/SlaughterhouseFiveOrTheChildrensCrusade/page/n41.

15. Oliver Pooley, "Substantivalist and Relationalist Approaches to Spacetime," *The Oxford Handbook of Philosophy of Physics*, ed. Robert Batterman (Oxford and New York: Oxford University Press, 2013), p. 522. For another clear description of the difference between relationalist and substantivalist theories of spacetime, see Field, "Can We Dispense With Space-Time?" 33.

16. Cf. Maudlin, "Buckets of Water and Waves of Space," 187. Noting that "[t]o defeat one tribe of relationists . . . is not to vanquish the whole nation," Maudlin entertains Newtonian [spatiotemporal] Relationalism—the view that "all spatiotemporal facts are facts about the relations between material bodies or, speaking four-dimensionally, material events." And, for the Newtonian Relationalist, the spatiotemporal relations would involve "a distance relation between noncontemporaneous events."

17. Edwin F. Taylor and John Archibald Wheeler, *Spacetime Physics* (San Francisco and London: W. H. Freeman and Company, 1966), p. 10. Cf. Friedman, *Foundations of Space-Time Theories*, pp. 34–35. Noting that "[t]he most important aim of our space-time theories is to describe the trajectories . . . of certain classes of physical particles," Friedman observes that one can "represent such a trajectory . . . by a *curve* [i.e., *world-line*] in space-time." Friedman then explains how one can represent "how much *and in what direction*" [emphasis added] a particular world-line is "'curved' or 'bent'" with respect to any given point on that world-line. Cf. Arntzenius, *Space, Time, and Stuff*, pp. 199–212. Arntzenius addresses the quantum field theories vis-à-vis the question of whether there is "an objective spacetime orientation (spacetime-handedness)."

18. D7.2 and other "spatiotemporal" analyses that follow include no variables that correspond to *times*: If a *spatiotemporal* relation obtains between two *spatiotemporal* entities, there would exist no *time* at which the *spatiotemporal* relation obtains.

19. In chapter 3, see D3.1, D3.2, and D3.9.

20. Graham Nerlich, "Space-Time Substantivalism," *The Oxford Handbook of Metaphysics*, p. 282.

21. See Brian Greene, *The Elegant Universe: Superstrings, Hidden Dimensions, and the Quest for the Ultimate Theory* (New York: W. W. Norton & Company, 1999), pp. 184–209.

22. Cf. Eagle, "Location and Perdurance," pp. 87–88. After noting that spacetime "*relationism*" is the view that "spacetime is not real, but is at best a useful fiction to adopt when representing genuine facts about the relations between material objects," Eagle observes that "[s]ince it is facts about material objects which are primary on this view, it would not be at all surprising if the best mereology of the locations of objects mirrored the best mereology for objects."

23. Brian Greene, *The Fabric of the Cosmos*, p. 416; see also Barbour, *The End of Time*, pp. 66–67. See also Abraham Pais, *Subtle is the Lord: The Science and the Life of Albert Einstein* (Oxford and New York: Oxford University Press, 2005), pp. 287–88. Pais reports that, for several years after Einstein published his work on General Relativity (1915), Einstein remained committed to Mach's Principle (the relativity of inertia), claiming in 1918 that it is among the three principles on which an adequate theory of gravitation should be grounded. Pais also reports that Einstein underscored his commitment several years later: "In 1922, Einstein noted that others were satisfied to proceed without [Mach's Principle] and added, 'This contentedness will appear incomprehensible to a later generation however.'" I thank Emory Kimbrough for this reference.

24. Cf. Friedman, *Foundations of Space-Time Theories*, p. 204. After observing that Einstein's "central philosophical motivation" for developing General Relativity was to formulate "a fully relational" concept of motion, Friedman claims that "[m]otion (and therefore spacetime) remains just as absolute in the relevant sense as it does in the Newtonian gravitation theory" and that "[t]he space-time structure used to define absolute motion . . . is not thereby eliminated in favor of relations between concrete physical objects and events."

25. Field, "Can We Dispense With Space-Time?" 42. Field entertains (and explains why he resists) various relationalist appeals to "heavy duty platonism" (i.e., the view that there exist "magnitude relations that relate aggregates of matter *to real numbers*"), the existence of "geometrically possible point-particles," and various modal solutions to "the problem of quantities."

26. John Earman, *World Enough and Space-Time: Absolute versus Relational Theories of Space and Time* (Cambridge, MA: The MIT Press, 1989), pp. 101–02; and see all of chapter 5 ("Relational Theories of Motion: A Twentieth-Century Perspective"), pp. 91–110. Earman claims (pp. 97–98) that Special Relativity and General Relativity allow "one to speak of *the* acceleration of a particle" in an absolute sense; that is, "without having to refer explicitly or implicitly to the motion of the particle in question with respect to other particles." Leaving open the possibility that

"there is some nonstandard interpretation of relativity theory" that operates without a concept of absolute acceleration, he argues that not even a nonstandard interpretation can do without a concept of absolute *rotation*, concluding that his argument "suffices to establish what classical absolutists desperately want to prove but never could, namely, that the very idea of space-time in its relativistic guise is irreconcilable with a full-blown relational conception of motion." Cf. Maudlin, "Buckets of Water and Waves of Space," p. 184.

27. Maudlin, "Buckets of Water and Waves of Space," p. 187.

28. Strictly, one who subscribes to the favored interpretation of Relativity would offer a slightly different account of the spinning bucket's absolute motion. According to the favored interpretation, it would not be accurate to say that the spatiotemporal directional relations that the bucket's left half bears to its right half are other than the relations that the left half will bear to the right after the bucket spins a quarter turn. Rather, according to the favored interpretation, the bucket is a *perduring* spatiotemporal entity—an entity that spatiotemporally persists in virtue of its being composed of a succession of spatiotemporal parts. Thus, to say that the perduring bucket undergoes absolute motion is to say that the spatiotemporal directional relations that obtain among the parts of any one of the perduring bucket's spatiotemporal parts are other than those that obtain among the parts of any one of the perduring bucket's other spatiotemporal parts.

29. A committed nominalist, Hartry Field would most certainly object to this directionalist/relationalist account of absolute motion. The account sketched here is explicated in terms of spatiotemporal locations, and the concept of a spatiotemporal location is analyzed in terms of spatiotemporal directional *relations*. I leave open the question of whether one can formulate directionalist/relationalist accounts of absolute motion and spatiotemporal location without commitment to the relations of Realism. See Field, "Can We Dispense With Space-Time?" pp. 43, 47 and also *Science without Numbers: A Defense of Nominalism* (Oxford: Oxford University Press, 2016).

30. See Earman, *World Enough and Space-Time*, p. 155.

31. Field, "Can We Dispense With Space-Time?" pp. 41, 40.

32. Arntzenius, *Space, Time, and Stuff*, pp. 139–40; Arntzenius addresses briefly the specification of a field's values if space is gunky—if it is not a dense array of points.

33. Field, "Can We Dispense With Space-Time?" p. 40.

34. Field, "Can We Dispense With Space-Time?" pp. 40–41. Cf. Arntzenius, *Space, Time, and Stuff*, pp. 171–72. Arntzenius is more skeptical than Field: After addressing relationalism, Special Relativity, and fields, Arntzenius concludes that the "prospects for relationalism seem just as bleak in the context of Special Relativity as they are in context of Newtonian physics."

35. Field, "Can We Dispense With Space-Time?" p. 41. See also Maudlin, "Buckets of Water and Waves of Space," p. 200.

36. Field, "Can We Dispense With Space-Time?" p. 41.

37. Maudlin, "Buckets of Water and Waves of Space," p. 201.

38. Cf. Earman, *World Enough and Space-Time*, p. 194: "Another more thoroughly relational and more constructivist approach would proceed in two steps: first,

the space-time manifold would be build up from physical events and their relations, and then with the manifold in hand the characterization of fields could proceed as usual."

39. Cf. Earman, *World Enough and Space-Time*, p. 159: "The antisubstantivalist can, of course, attempt to dispense with some or all of this apparatus in favor of another means of specifying a relationally pure state of affairs and view all the above as merely giving representations of the underlying relational state." Earman adds, "While not prejudging the success of such an endeavor, . . . the burden of proof rests with the antisubstantivalist."

40. Cf. Jessica Wilson, "Newtonian Forces," *British Journal for Philosophy of Science* 58 (2007), 173–205. Although Wilson argues that Newtonian forces exist, siding with the view that such forces are "aspects . . . of the nonforce entities necessitating them" (p. 200), she does leave open the question regarding their ontological status— "[A]re they manifested dispositions, causal relations, sui generis?" (p. 203). See also Wilson's entry in *The Encyclopedia of Philosophy*, 2nd ed., s.v. "Force [Addendum]" (Detroit: Thomson Gale, 2006), pp. 690–92.

41. Brian Greene, "Listening to the Big Bang," *Smithsonian*, May 2014, p. 19.

42. Taylor and Wheeler, *Spacetime Physics*, p. 175. See Maudlin, *Philosophy of Physics*, pp. 131–40. See also Pooley, "Substantivalist and Relationalist Approaches to Spacetime," pp. 538–39: In explaining that, per General Relativity, "gravitational phenomena are not understood as resulting from the action of forces," Pooley notes that "[t]he 'force' that holds us on the surface of the Earth and the 'force' pinning the astronaut to the floor of the accelerating rocket ship are literally of one and the same kind."

43. Taylor and Wheeler, *Spacetime Physics*, pp. 191–92. Cf. Jim Al-Khalili, *Paradox: The Nine Greatest Enigmas in Physics* (New York: Broadway Paperback, an imprint of Random House, 2012), p. 53: "Einstein's General Theory of Relativity . . . says that gravity isn't really a force as such—like an invisible rubber band that pulls all matter together—but rather a measure of the shape of space itself around all masses."

44. Greene, "Listening to the Big Bang," p. 20.

45. Greene, *The Fabric of the Cosmos*, p. 69; see also p. 418.

46. John Archibald Wheeler with Kenneth W. Ford, *Geons, Black Holes and Quantum Foam: A Life in Physics* (New York: W. W. Norton, 2000), p. 235. I thank Emory Kimbrough for this reference.

47. See Chris Impey, *Humble Before the Void: A Western Astronomer, His Journey East, and a Remarkable Encounter between Western Science and Tibetan Buddhism* (West Conshohocken, PA: Templeton Press, 2014), p. 174. After citing the Wheeler quotation, Impey notes that "[t]he most common analogy is a rubber sheet that bends when a mass is present": "Objects rolling across the rubber sheet are deflected by the curved space."

48. Stephen Hawking, *The Universe in a Nutshell* (New York: Bantam Books, 2001), pp. 19–21. Cf. Lawrence Krauss, *A Universe from Nothing: Why There is Something Rather than Nothing* (New York: Atria Paperback, an imprint of Simon & Schuster, Inc., 2012), p. 26.

49. See Tilman Sauer, "A Brief History of Gravitational Lensing," *Einstein Online* 4 (2010), 1005. (I thank Emory Kimbrough for this citation.) See also Hawking, *The Universe in a Nutshell*, pp. 19–21; William Keel, *The Sky at Einstein's Feet*, Springer-Praxis Books in Popular Astronomy (New York: Springer-Verlag, 2006), pp. 97–119.

50. Greene, "Listening to the Big Bang," pp. 20, 22–23; *The Fabric of the Cosmos*, pp. 282–94. See also Keel, *The Sky at Einstein's Feet*, pp. 205–35; Tim Maudlin, *The Metaphysics Within Physics* (Oxford: Oxford University Press, 2007), pp. 40–44.

51. Al-Khalili, *Paradox: The Nine Greatest Enigmas in Physics*, p. 56.

52. Greene, "Listening to the Big Bang," p. 20.

53. See Impey, *Humble Before the Void*, pp. 93, 97. Astronomer Impey compares the cosmic expansion of galaxies to the expansion of beads glued to the surface of an balloon: As the balloon expands, the beads move further apart, but the beads themselves do not grow larger as the balloon expands. He also uses the example of raisins moving further apart in "a raisin loaf baking in an oven."

54. Greene, *The Fabric of the Cosmos*, p. 419. See also Greene, "Listening to the Big Bang," p. 24: "[S]pace itself should be subject to quantum jitters too [which means that] space should undulate like the surface of a boiling pot of water."

55. Greene, "The Detection of Gravitational Waves Was a Scientific Breakthrough, but What's Next?" *Smithsonian*, April 2016. https://www.smithsonianmag.com/science-nature/detection-gravitational-waves-breakthrough-whats-next-180958511/.

56. Greene, *The Fabric of the Cosmos*, pp. 421–42. See also Greene, "The Detection of Gravitational Waves Was a Scientific Breakthrough, but What's Next?"

57. Greene, *The Fabric of the Cosmos*, p. 420.

58. Pooley, "Substantivalist and Relationalist Approaches to Spacetime," p. 539.

59. Nerlich, "Space-Time Substantivalism," p. 282.

60. Cf. Brentano, *Psychology from an Empirical Standpoint*, p. 362: "Newton was certainly correct in maintaining that every spatial point as such must be specifically located."

61. Cf. Maudlin, "Buckets of Water and Waves of Space," 202. Noting that "[t]he deep ontology of physics twenty years hence may be as different from today's as today's is from that of Democritus," Maudlin speculates that "[t]he pendulum may swing back to relationism or, more likely, the structures postulated may have such unfamiliar properties that the notions of substance and relation cease to have any clear application."

62. Jeremy I. Pfeffer and Schlomo Nir, *Modern Physics: An Introductory Text* (London: Imperial College Press, 2000), p. 81.

63. This thought experiment may not be too farfetched given that there are string theorists who endorse the "multiverse" view—the view that there exist multiple independent spacetime manifolds. See Brian Green, *The Hidden Reality: Parallel Universes and the Deep Laws of the Cosmos* (New York: Vintage Books, an imprint of Random House, 2011).

64. Isaac Newton's fourth letter to Richard Bentley. http://www.newtonproject.ox.ac.uk/view/texts/normalized/THEM00258.

65. For an accessible introduction to the Special Relativity Objection to Presentism, see Ney, *Metaphysics: An Introduction*, pp. 140–42. Cf. Sider, *Four-Dimensionalism*, pp. 79–87; D. H. Mellor, *Real Time II*, International Library of Philosophy (Abingdon and New York: Routledge, 1998), pp. 56–57; Michael C. Rea, "Four-Dimensionalism," *The Oxford Handbook of Metaphysics*, pp. 246–80.

66. Markosian, "A Defense of Presentism," pp. 173–74. Markosian uses 'STR' to refer to Special [Theory of] Relativity; to avoid confusion with my use of 'STR' (to refer to Spatiotemporal Relationalism), I replace Markosian's use of 'STR' with 'SRL' in my explication of his work.

67. Markosian, "A Defense of Presentism," p. 175.

68. Whether Special Relativity requires rejecting a unique present is not obvious. See Zimmerman, "The Privileged Present: Defending an 'A-theory' of Time," pp. 219–21.

69. Cf. Arntzenius, *Space, Time, and Stuff*, p. 27: "Using intuitions about what is and what is not metaphysically possible in order to decide which theories of physics are right, strikes me as putting the cart before the horse. It seems to me a better tactic to first figure out what our best theory is, and then use that theory in determining what the metaphysical possibilities are. . . . [A]rguments which start with *prima facie* intuitions as to what is and what is not metaphysically possible, and end with conclusions as to what the best theories in science must be like, do not make for sound methodology, and would hamper scientific progress."

70. Emory Kimbrough tells me that there is reason to be hopeful that there will emerge a nonmysterious alternative that is consistent with relationalism/directionalism—a quantum theory of gravity that requires no mysterious action-at-a-distance force *or* warped substantivalist spacetime. See Rickles, *The Philosophy of Physics*, pp. 185–89 and Greene, *The Elegant Universe*, pp. 120–23.

71. See Jessica Wilson, "The Question of Metaphysics," *The Philosophers' Magazine* (Summer 2016), p. 94. As this chapter suggests, I am sympathetic with Jessica Wilson's defense of what she calls "the 'embedded' conception" of metaphysics—the view that "metaphysics is embedded in other disciplines . . . and all other areas of investigation relevant to the subject matter in question." Wilson argues that "metaphysical notions and posits are embedded in the notions and posits of other areas of investigation" (e.g., psychology, neurology, mathematics, and "the chemical and material sciences") and that "the directions of potential influence [go] both to and from metaphysics."

In writing about his traveling to India to teach physics, astronomy, and cosmology to Tibetan monks, astronomer Chris Impey reports that the monks asked, "'What is the nature of space?'" and also "If invisible space didn't have stars and galaxies in it to mark it, could we measure it at all? What is space if it's really nothing, and how can nothing expand? Is space quantized like matter, or can it be infinitely subdivided? Is it a phenomenon or just an abstraction?" After noting that "[t]hese are among the great metaphysical questions of the cosmos," Impey writes, "I will have no answers for my students." See Impey, *Humble Before the Void*, pp. 43–44. If asked *why* he lacks answers to these "great metaphysical questions," what answer should Impey offer? My hope is that Impey will *not* claim that, as a scientist, he has nothing relevant to say about the metaphysics of space or spacetime. Rather, if what physicists and astronomers do know about motion, fields, gravity, and the expansion of galaxies is best explained by positing substantivalist spacetime, then this would count as an abductive argument for Spatiotemporal Substantivalism—an argument that the metaphysician should take seriously when forging a philosophical theory of location. And, if the metaphysician develops compelling philosophical objections to substantivalist spacetime, then further work is in order: Can the metaphysician, physicist, and astronomer together forge a reductivist theory of spacetime that *does* preserve what the scientists know about motion, fields, gravity, and galaxies? Or can they together explicate Spatiotemporal Substantivalism in such a way that the metaphysician's objections cease to be compelling?

72. Maudlin, *Philosophy of Physics*, p. 66.

73. Cf. Greene, "Relativity and the Absolute: Is Spacetime an Einsteinian Abstraction or a Physical Entity?" p. 75. After noting that space and then spacetime were invoked to define accelerated motion with respect to Newton's physics and then Special Relativity, Greene emphasizes that, unlike the absolute space of which Newton conceived, the spacetime of Special Relativity is not absolute but can instead warp and curve. See also Earman, *World Enough and Space-Time*, p. 155: "When relativity theory banished the ether, the space-time manifold M began to function as a kind of dematerialized ether needed to support the fields" (e.g., the electromagnetic field).

Bibliography

Abbott, Edwin A. *Flatland: A Romance of Many Dimensions*. 2nd ed. New York: Dover Publications, 1952.

Al-Khalili, Jim. *Paradox: The Nine Greatest Enigmas in Physics*. New York: Broadway Paperback, an imprint of Random House, 2012.

Aristotle. *Physics*. Translated by R. P. Hardie and R. K. Gayle. *The Basic Works of Aristotle*. Edited by Richard McKeon. New York: Random House, 1941.

———. *Posterior Analytics*. Translated by A. J. Jenkinson. *The Basic Works of Aristotle*. Edited by Richard McKeon. New York: Random House, 1941.

Arntzenius, Frank. *Space, Time, and Stuff*. Oxford and New York: Oxford University Press, 2012.

Banchoff, Thomas. *Beyond the Third Dimension*. Scientific American Library Series. New York: W. H. Freeman & Co., 1990.

Barbour, Julian. *The End of Time: The Next Revolution in Physics*. Oxford and New York: Oxford University Press, 1999.

Berkeley, George. *Philosophical Commentaries*. Notebooks B and A. Edited by A. A. Luce. *The Works of George Berkeley, Bishop of Cloyne*. Edited by A. A. Luce and T. E. Jessop. Vol. I. London: Thomas Nelson and Sons Ltd., 1951: 1–139.

———. *A Treatise Concerning the Principles of Human Knowledge*. Edited by T. E. Jessop. *The Works of George Berkeley, Bishop of Cloyne*. Edited by A. A. Luce and T. E. Jessop. Vol. II. London: Thomas Nelson and Sons Ltd., 1949: 19–113.

Bourne, Craig. *A Future for Presentism*. Oxford: Clarendon Press, an imprint of Oxford University Press, 2006.

Brentano, Franz. "On *Ens Rationis*." *Psychology from an Empirical Standpoint*. Edited by Oscar Kraus. English edition edited by Linda L. McAlister. Translated by Antos C. Rancurello, D. B. Terrell, and Linda L. McAlister. New York: Humanities Press, 1973: 339–68.

———. "On What is Continuous." Dictated 22 November 1914. *Space, Time and the Continuum*. Translated by Barry Smith. London, New York, and Sydney: Croom Helm, 1988: 1–44.

———. *Psychology from an Empirical Standpoint*. Edited by Oscar Kraus. English edition edited by Linda L. McAlister. Translated by Antos C. Rancurello, D. B. Terrell, and Linda L. McAlister. New York: Humanities Press, 1973.

———. *The Theory of Categories*. Translated by Roderick M. Chisholm and Norbert Guterman. Vol. 8, *Melbourne International Philosophy Series*. Edited by Jan T. J. Srzednicki. The Hague, Boston, and London: Martinus Nijhoff Publishers, 1981.

———. "What We Can Learn about Space and Time from the Conflicting Errors of the Philosophers." Dictated 23 February 1917. *Space, Time and the Continuum*. Translated by Barry Smith. London, New York, and Sydney: Croom Helm, 1988: 156–81.

Carroll, John W. and Ned Markosian. *An Introduction to Metaphysics*. Cambridge and New York: Cambridge University Press, 2010.

Carter, William R. "In Defense of Undetached Parts." *Pacific Philosophical Quarterly*. 64 (1983): 126–43.

Carter, William R. and H. Scott Hestevold. "On Passage and Persistence." *American Philosophical Quarterly*. 31 (1994): 269–83.

Cartwright, Richard. "Scattered Objects." *Analysis and Metaphysics: Essays in Honor of R.M. Chisholm*. Edited by Keith Lehrer. Dordrecht and Boston: D. Reidel Publishing Co., 1975.

Casati, Roberto and Achille C. Varzi. *Holes and Other Superficialities*. Cambridge, MA, and London: Bradford Books, an imprint of The MIT Press, 1994.

———. *Parts and Places: The Structures of Spatial Representation*. A Bradford Book. Cambridge, MA, and London: The MIT Press, 1999.

Chisholm, Roderick M. "Boundaries." *On Metaphysics*. Minneapolis: The University of Minnesota Press, 1989: 83–89.

———. "Events and Propositions." *Noûs*. IV (1970): 15–24.

———. "Identity Criteria for Properties." *The Harvard Review of Philosophy*. 2 (1992): 14–16.

———. "Mereological Essentialism: Some Further Considerations." *The Review of Metaphysics*. 28 (1975): 477–84.

———. "Parts as Essential to Their Wholes." *The Review of Metaphysics*. 26 (1973): 581–603.

———. *Person and Object: A Metaphysical Study*. LaSalle, IL: Open Court Publishing Co., 1976.

———. "Properties and States of Affairs." *On Metaphysics*. Minneapolis: The University of Minnesota Press, 1989: 141–49.

———. *A Realistic Theory of Categories*. Cambridge: Cambridge University Press, 1996.

———. "Scattered Objects." *On Metaphysics*. Minneapolis: The University of Minnesota Press, 1989: 90–95.

———. "States and Events." *On Metaphysics*. Minneapolis: The University of Minnesota Press, 1989: 150–55.

Dasgupta, Shamik. "Substantivalism vs Relationalism About Space in Classical Physics." *Philosophy Compass*. 10/9 (2015): 601–24.

Donnelly, Maureen. "Parthood and Multi-Location." *Oxford Studies in Metaphysics*, Vol. 5. Edited by Dean Zimmerman. Oxford: Oxford University Press, 2010: 203–43.
Eagle, Antony. "Location and Perdurance." *Oxford Studies in Metaphysics*, Vol. 5. Edited by Dean Zimmerman. Oxford: Oxford University Press, 2010: 53–94.
Earman, John. *World Enough and Space-Time: Absolute versus Relational Theories of Space and Time*. Cambridge, MA: The MIT Press, 1989.
Field, Hartry. "Can We Dispense With Space-Time?" *PSA: Proceedings of the Biennial Meeting of the Philosophy of Science Association*. 1984 (1984): 33–90.
———. *Science without Numbers: A Defense of Nominalism*. Oxford: Oxford University Press, 2016.
Forrest, Peter. "Conflicting Intuitions About Space." *Mereology and Location*. Edited by Shieva Kleinschmidt. Oxford and New York: Oxford University Press, 2014: 117–31.
Friedman, Michael. *Foundations of Space-Time Theories: Relativistic Physics and Philosophy of Science*. Princeton, NJ: Princeton University Press, 1983.
Goodman, Nelson. *The Structure of Appearance*. Cambridge, MA: Harvard University Press, 1951.
Greene, Brian. "The Detection of Gravitational Waves Was a Scientific Breakthrough, but What's Next?" *Smithsonian*, April 2016. https://www.smithsonianmag.com/science-nature/detection-gravitational-waves-breakthrough-whats-next-180958511/.
———. *The Elegant Universe: Superstrings, Hidden Dimensions, and the Quest for the Ultimate Theory*. New York: W. W. Norton & Company, 1999.
———. *The Fabric of the Cosmos: Space, Time, and the Texture of Reality*. New York: Vintage Books, 2005.
———. *The Hidden Reality: Parallel Universes and the Deep Laws of the Cosmos*. New York: Vintage Books, an imprint of Random House, 2011.
———. "Listening to the Big Bang." *Smithsonian*, May 2014: 19–26.
Harte, Verity. *Plato on Parts and Wholes: The Metaphysics of Structure*. Oxford: Clarendon Press, an imprint of Oxford University Press, 2002.
Haslanger, Sally. "Persistence Through Time, Substantivalism." *The Oxford Handbook of Metaphysics*. Edited by Michael J. Loux and Dean W. Zimmerman. Oxford and New York: Oxford University Press, 2003: 315–54.
Hawking, Stephen. *The Universe in a Nutshell*. New York: Bantam Books, a division of Random House, Inc. 2001.
Hawley, Katherine. *How Things Persist*. Oxford: Clarendon Press, an imprint of Oxford University Press, 2001.
Herbert, Gary B. "Hobbes's Phenomenology of Space." *Journal of the History of Ideas*. 48 (1987): 709–17.
Hestevold, H. Scott. "Boundaries, Surfaces, and Continuous Wholes." *The Southern Journal of Philosophy*. XXIV (1986): 235–45.
———. "Conjoining." *Philosophy and Phenomenological Research*. 41 (1981): 371–85.

———. "A Metaphysical Study of Aggregates and Continuous Wholes." PhD diss., Brown University, 1978.

———. "Presentism: Through Thick and Thin." *Pacific Philosophical Quarterly*. 89 (2008): 325–47.

Hobbes, Thomas. *Elements of Philosophy: Concerning the Body*. Vol. I, *The English Works of Thomas Hobbes of Malmesbury*. Edited by Sir William Molesworth. London: John Bohn, 1839.

Hoffman, Joshua and Gary Rosenkrantz. "Are Souls Unintelligible?" *Philosophical Perspectives*. Vol. 5, Philosophy of Religion. Atascadero, CA: Ridgeview Publishing Co., 1991: 183–212.

Horwich, Paul. *Assymetries in Time: Problems in the Philosophy of Science*. Cambridge, MA, and London: Bradford Books, an imprint of The MIT Press, 1987.

Hudson, Hud. *The Metaphysics of Hyperspace*. Oxford and New York: Oxford University Press, 2005.

Impey, Chris. *Humble Before the Void: A Western Astronomer, His Journey East, and a Remarkable Encounter between Western Science and Tibetan Buddhism*. West Conshohocken, PA: Templeton Press, 2014.

Johnson, Jr. Major L. "Events as Recurrables." *Analysis and Metaphysics: Essays in Honor of R.M. Chisholm*. Edited by Keith Lehrer. Dordrecht and Boston: D. Reidel Publishing Co., 1975: 209–26.

Keel, William. *The Sky at Einstein's Feet*. Springer-Praxis Books in Popular Astronomy. New York: Springer–Verlag, 2006.

Koslicki, Kathrin. *The Structure of Objects*. Oxford and New York: Oxford University Press, 2008.

Krauss, Lawrence. *A Universe from Nothing: Why There is Something Rather than Nothing*. New York: Atria Paperback, an imprint of Simon & Schuster, 2012.

Kutach, Douglas. "The Asymmetry of Influence." *The Oxford Handbook of Philosophy of Time*. Edited by Craig Callender. Oxford: Oxford University Press, 2011: 247–75.

Leibniz, G. W. *Monadology*. Edited by G. H. R. Parkinson. Translated by Mary Morris and G. H. R. Parkinson. London: J. M. Dent & Sons Ltd, 1973: 179–94.

Leibniz, G. W. and Samuel Clarke. "Correspondence." *The Leibniz-Clarke Correspondence, Together with Extracts from Newton's "Principia" and "Opticks"*. Edited by H. G. Alexander. Manchester: Manchester University Press, 1956.

Le Poidevin, Robin. *Travels in Four Dimensions*. Oxford and New York: Oxford University Press, 2003.

Lewis, David. *On the Plurality of Worlds*. Oxford: Blackwell Publishing, 1986.

———. *Parts of Classes*. Oxford and Cambridge, MA: Basil Blackwell, 1991.

Lewis, David and Stephanie. "Holes." *Australasian Journal of Philosophy*. 48 (1970): 206–12.

Lombard, Lawrence. *Events: A Metaphysical Study*. London, Boston, and Henley: Routledge & Kegan Paul, 1986.

Loux, Michael J. and Dean W. Zimmerman. "Introduction." *The Oxford Handbook of Metaphysics*. Edited by Loux and Zimmerman. Oxford and New York: Oxford University Press, 2003: 1–7.

Markosian, Ned. "Brutal Composition." *Philosophical Studies*. 92 (1998): 211–49.

———. "A Defense of Presentism." *Oxford Studies in Metaphysics*. Vol. 1. Oxford: Clarendon Press, an imprint of Oxford University Press, 2004: 47–82.

———. "The Right Stuff." *Australasian Journal of Philosophy*. 93 (2015): 665–87.

———. "Simples." *Australasian Journal of Philosophy*. 76 (1998): 213–26.

———. "Simples, Stuff, and Simple People." *The Monist*. 87 (2004): 405–28.

———. "A Spatial Approach to Mereology." *Mereology and Location*. Edited by Shieva Kleinschmidt. Oxford and New York: Oxford University Press, 2014: 69–90.

———. "What Are Physical Objects?" *Philosophy and Phenomenological Research*. 61 (2000): 375–95.

Maudlin, Tim. "Buckets of Water and Waves of Space: Why Spacetime is Probably a Substance." *Philosophy of Science*. 60 (1993): 183–203.

———. *The Metaphysics Within Physics: Space and Time*. Oxford: Oxford University Press, 2007.

———. *Philosophy of Physics: Space and Time*. Princeton, NJ, and Oxford: Princeton University Press, 2012.

McGrath, Matthew. "No Objects, no Problem?" *Australasian Journal of Philosophy*. 83 (2005): 457–86.

Mellor, D. H. *Real Time II*. International Library of Philosophy. Abingdon and New York: Routledge, 1998.

Merricks, Trenton. *Objects and Persons*. Oxford and New York: Oxford University Press, 2003.

Moore, George Edward. *Some Main Problems of Philosophy*. London: George Allen & Unwin; New York: Humanities Press, 1956.

Nerlich, Graham. "Space-Time Substantivalism." *The Oxford Handbook of Metaphysics*. Edited by Michael J. Loux and Dean W. Zimmerman. Oxford and New York: Oxford University Press, 2003: 281–314.

Newton, Isaac. *De Gravitatione*. In *Philosophical Writings*. Edited by Andrew Janiak. Cambridge Texts in the History of Philosophy. Series edited by Karl Ameriks and Desmond M. Clarke. Cambridge, New York, Melbourne, Madrid, Cape Town, Singapore, São Paulo: Cambridge University Press, 2004. https://www.hrstud.unizg.hr/_download/repository/Newton,_Philosophical_Writings.pdf.

———. Fourth letter to Richard Bentley. http://www.newtonproject.ox.ac.uk/view/texts/normalized/THEM00258.

———. *Sir Isaac Newton's Mathematical Principles of Natural Philosophy and his System of the World*. Translated by Andrew Motte. Revised by Florian Cajori. Berkeley: University of California Press, 1934.

Ney, Alyssa. *Metaphysics: An Introduction*. London and New York: Routledge, an imprint of Taylor & Francis Group, 2014.

Nolan, Daniel. "Balls and All." *Mereology and Location*. Edited by Shieva Kleinschmidt. Oxford and New York: Oxford University Press, 2014: 91–116.

North, Jill. "Time in Thermodynamics." *The Oxford Handbook of Philosophy of Time*. Edited by Craig Callender. Oxford: Oxford University Press, 2011: 312–50.

Oaklander, L. Nathan. *The Ontology of Time*. Amherst, NY: Prometheus Books, 2004.

Olson, Eric T. *What Are We? A Study in Personal Ontology*. New York: Oxford University Press, 2007.

Pais, Abraham. *Subtle is the Lord: The Science and the Life of Albert Einstein*. Oxford and New York: Oxford University Press, 2005.

Pfeffer, Jeremy I. and Schlomo Nir. *Modern Physics: An Introductory Text*. London: Imperial College Press, 2000.

Poincaré, Henri. *Science and Method*. Translated by Francis Maitland. London, Edinburgh, Dublin, & New York: Thomas Nelson and Sons, 1914.

Pooley, Oliver. "Substantivalist and Relationalist Approaches to Spacetime." *The Oxford Handbook of Philosophy of Physics*. Edited by Robert Batterman. Oxford and New York: Oxford University Press, 2013: 522–86.

Price, Huw. "The Flow of Time." *The Oxford Handbook of Philosophy of Time*. Edited by Craig Callender. Oxford: Oxford University Press, 2011: 276–311.

Quine, Willard van Orman. *Word and Object*. Cambridge, MA: The MIT Press, 1960.

Rea, Michael. "Four-Dimensionalism." *The Oxford Handbook of Metaphysics*. Edited by Michael J. Loux and Dean W. Zimmerman. Oxford and New York: Oxford University Press, 2003: 246–80.

Reid, Thomas. *Essays on the Intellectual Powers of Man*. Edited by Ronald E. Beanblossom and Keith Lehrer. Indianapolis, IN: Hackett Publishing Co., 1983.

Rickles, Dean. *The Philosophy of Physics*. Cambridge, UK, and Malden, MA: Polity Press, 2016.

Sauer, Tilman. "A Brief History of Gravitational Lensing." *Einstein Online*. 4 (2010): 1005. http://www.einstein-online.info/spotlights/grav_lensing_history.1.html.

Sider, Theodore. "Against Parthood." *Oxford Studies in Metaphysics*. Edited by Karen Bennett and Dean W. Zimmerman. Vol. 8. Oxford and New York: Oxford University Press, 2013: 237–93.

———. "'Bare Particulars.'" *Philosophical Perspectives*. 20 (2006): 387–97.

———. *Four-Dimensionalism: An Ontology of Persistence and Time*. Oxford: Clarendon Press, an imprint of Oxford University Press, 2001.

———. "Van Inwagen and the Possibility of Gunk." *Analysis*. 53 (1993): 285–89.

———. *Writing the Book of the World*. Oxford: Oxford University Press, 2011.

Simons, Peter. "Where It's At: Modes of Occupation and Kinds of Occupant." *Mereology and Location*. Edited by Shieva Kleinschmidt. Oxford and New York: Oxford University Press, 2014: 59–68.

Sklar, Lawrence. *Space, Time, and Spacetime*. Berkeley, Los Angeles, London: University of California Press, 1977.

Taylor, Edwin F. and John Archibald Wheeler. *Spacetime Physics*. San Francisco and London: W. H. Freeman and Company, 1966.

Thomasson, Amie L. *Ordinary Objects*. Oxford: Oxford University Press, 2007.

Unger, Peter. "There are no Ordinary Things." *Synthese*. 41 (1979): 117–54.

Uzquiano, Gabriel. "Plurals and Simples." *The Monist*. 87 (2004): 429–51.

van Cleve, James. "The Moon and Sixpence: A Defense of Mereological Universalism." *Contemporary Debates in Metaphysics*. Edited by John Hawthorne, Theodore Sider, and Dean Zimmerman. Oxford: Blackwell, 2008: 321–40.
van Inwagen, Peter. "The Doctrine of Arbitrary Undetached Parts." *Pacific Philosophical Quarterly*. 62 (1981): 123–37.
———. *Material Beings*. Ithaca, NY: Cornell University Press, 1990.
———. *Metaphysics*. 4th ed. Boulder, CO: Westview Press, 2015.
Vonnegut, Kurt. *Slaughterhouse Five*. 1969. https://archive.org/details/SlaughterhouseFiveOrTheChildrensCrusade/page/n41.
Wheeler, John Archibald with Kenneth W. Ford. *Geons, Black Holes and Quantum Foam: A Life in Physics*. New York: W. W. Norton, 2000.
Wilson, Jessica. "Force [Addendum]." *The Encyclopedia of Philosophy*. 2nd ed. Detroit: Thomson Gale, 2006: 690–92.
———. "Newtonian Forces." *British Journal for Philosophy of Science*. 58 (2007): 173–205.
———. "The Question of Metaphysics." *The Philosophers' Magazine* (Summer 2016): 90–96.
Zimmerman, Dean W. "Could Extended Objects Be Made Out of Simple Parts? An Argument for 'Atomless Gunk'." *Philosophy and Phenomenological Research*. 56 (1996): 1–29.
———. "Presentism and the Space-Time Manifold." *The Oxford Handbook of Philosophy of Time*. Edited by Craig Callender. Oxford: Oxford University Press, 2011: 163–244.
———. "The Privileged Present: Defending an 'A-Theory' of Time." *Contemporary Debates in Metaphysics*. Edited by Theodore Sider, John Hawthorne, and Dean W. Zimmerman. Malden, MA: Blackwell, 2007: 212–25.

Index

Page references for figures are italicized.

Abbott, Edwin A., 40–41
Absolute Motion Argument for substantivalist space (AM), 63–71
acceleration, 145, 150–51, 185, 189
Aristotle, 63, 81n36, 120
Arntzenius, Frank, xiiin2, 13n4, 30n3, 135n13, 182nn4–5, 183n17, 185nn32, 34, 188n69
atom, mereological. *See* simple, mereological
axioms regarding spatial directional relations, 22, 25–28, *27*

Banchoff, Thomas, 47n14
Barbour, Julian, 79n22, 80n31, 184n23
Berkeley, George, 12n1, 39–40, 68–69, 77, 79n23, 80nn29, 32
Big Bang, 159, 163–64, 173, 176–77, 179
black holes, 158, 161, 163, 176–77
boundaries, x–xi, xiiin1, 8–9, 11–12, 34, 78, 85, 100–102, 105, 108, 111n31, 113–30, *123*; Chisholm/Brentano theory of, 115–17, 120–22; directionalist theory of, 122–30; Hestevold's Brentano-esque theory of, 114–15, 117–22; of scattered objects, 121, 127–30

Bourne, Craig, 30n3
Brentano, Franz, x–xi, xiiin1, 8–9, 11, 14n13, 54, 60, 79n22, 109n6, 114–22, 124–26, 135nn16, 18, 181n2, 187n60

Carroll, John W., 110n28, 110n30
Carter, William R., 14n16, 182n9, 183n12
Cartwright, Richard, 4, 14n18, 33, 46n3, 90–91, 110n15
Casati, Roberto and Achille C. Varzi, 13n5, 14n15, 131
Chisholm, Roderick M., x, xiiin1, 14nn13, 18, 30nn1–2, 46n6, 75, 80n35, 91, 93, 109nn5, 10, 110nn15, 17, 115–17, 119–21, 125, 134n3, 135nn7, 17
Clarke, Samuel, 49–50, 52, 54, 56, 58–59, 67–69, 79nn13, 15
composition, ix–xi, 7–8, 10–12, 36–37, 46n3, 128–30, 140–42, 147. *See also* Special Composition Question (SCQ)
Conjoining (CON), xi, 8, 14n12, 83–84, 108; formulation and implications of, 84–91, 93–96, 109nn9, 12; objections to, 91–103

contact, x, xiiinn1–2, 10, 14n13, 86, 88, 99, 104–5, 110n20, 111n29, 135n7
continuous objects. *See* materially solid objects

Dasgupta, Shamik, 14n10, 78nn3, 6, 79n24
dependent particulars, 8–9, 11, 78, 101–2, 114–16, 120
dimension, spatial or spatiotemporal, 7, 40, 43–46, 76, 145, 149
Directionalist Spatiotemporal Relationalism (DSR), 9, 138, 148–49, 169–81
Directionalist Theory of Space (DTS), 6–8, 10–12, 17–18, 22, 33–37, 39–40, 42–43, 46, 47nn3–4, 53–54, 76–78, 133–34, 137–38, 142–44, 149; objections to, 40–43, 49–76
Donnelly, Maureen, 182n3

Eagle, Antony, 182n3, 184n22
Earman, John, 150, 152, 184n26, 185nn30, 38, 186n39, 189n73
Einstein, Albert, 138–39, 150, 156–60, 163, 173, 183n13, 184nn23–24, 186n43
electromagnetic field. *See* fields
Endurance, 141, 170, 183nn11–12
expansion of the universe, 9, 158–59, 162, 164, 170, 173–75, 178–80, *180*, 187n53, 189n71
Eternalism vis-à-vis time and spacetime, 21–22, 140–42, 145, 148, 166, 179, 180
events, 12, 18, 21, 30nn1–3, 47n8, 138–39, 158, 164, 175–76, 183n16, 184n24, 186n38. *See also* states of affairs

Field, Hartry, 47n9, 72, 150, 152–53, 183n15, 184n25, 185nn29, 34–35
fields, 9, 72, 151–55, 162, 184n17, 185nn32, 34, 186n38, 189nn71, 73
field theory. *See* fields
Flatland, 40–44

force, 7, 41, 63–66, 68–71, 79n23, 80n32, 105–7, 142, 150, 155–58, 165, 169, 173–74, 186nn40, 42–43. *See also* gravity
Forrest, Peter, 13n5
Friedman, Michael, 47n8, 79n12, 138, 183nn13, 17, 184n24

God, 40, 50, 54, 56–58, 78n5, 79nn13, 15, 116
Goodman, Nelson, 109n6
gravitational waves, 9, 159–64, 168–70, 174, *180*, 180–81
gravity, 9, 106, 155–58, 160, 168–70, 173–75, 186n43, 188n70, 189n71
Greene, Brian, 47nn16–17, 78n1, 80n27, 138, 150, 156–61, 182n8, 184n21, 186n45, 187nn50, 54, 56, 63, 188n70, 189n73
gunk, 13n4, 111n33, 185n32

Harte, Verity, 108n2, 135n11
Haslanger, Sally, 183n11
Hawking, Stephen, 157–58, 186n48, 187n49
Hawley, Katherine, 183n11
Herbert, Gary B., 13n2
Hestevold, Erik, xii
Hestevold, H. Scott, 14nn12–13, 102, 117–21, 134n1, 182n9, 183n12
Hobbes, Thomas, 4, 13n2, 28, 33
Hoffman, Joshua, 15n20
holes, x–xi, 2, 8–10, 14nn14–15, 78, 91, 108, 113–14, 130–33, 136n20, 137, 149
Horwich, Paul, 30n3
Hudson, Hud, 41–43
hyperspace, 13n5, 40–43, 46, 76

identity criterion for spatial directional relations, 7–8, 71–76
Impey, Chris, 186n47, 187n53, 188n71
inverted world, 3, 5–7, 54–60, 71, 148
Inverted World Argument (IW), 54–60, *59*, 78n9

Johnson, Jr. Major L., 30n2

Keel, William, 187n49–50
Al-Khalili, Jim. 158, 186n43
Kimbrough, Emory, xii–xiii, 80n31, 184n23, 186n46, 187n49, 188n70
Koslicki, Kathrin, 109n7, 110n19, 135nn11, 14
Krauss, Lawrence, 187n48
Kutach, Douglas, 30n3

Leibniz, G. W., ix–xii, 4–7, 11, 15n21, 17, 29, 33, 38, 40, 47n8, 49, 52–54, 56–58, 62, 65, 67–72, 76, 78nn1, 3, 9–10, 79n12, 80n27, 134, 135n16, 138, 144–45, 154, 175, 181, 183n13
Le Poidevin, Robin, 13n3
Lewis, David, 14n14, 109n6, 130–31, 133, 136n20, 182n4
Lewis, Stephanie, 14n14, 130–31, 133, 136n20
LIGO (Laser Interferometer Gravitational-Wave Observatory), 161
location, x–xii, 1–2, 4–7, 9, 17–19, 33–37, 38–40, 46n4, 54, 65, 71–72, 76, 89–90, 95, 102, 114, 127–30, 132–34, 137–38, 141, 143–44, 146–47, 150–51, 154–55, 161–63, 166, 175–77, 179, 181, 185n29, 189n71
Lombard, Lawrence, 30n2
Loux, Michael J., 14n9

Mach, Ernst, 80n31, 150, 161, 175, 183n13, 184n23
Markosian, Ned, 4, 14nn16–17, 15n21, 46n5, 108n3, 110nn28, 30, 112n34, 171–72, 182n10, 188n66
materially solid objects, x–xi, 8–11, 147; analysis of the concept of, 90; and the bearing, exhibiting, or obtaining of spatial directional relations, 20, 22–29, *23*, *24*, 28–29, 35–37; and boundaries, 113–30; composition and, 83, 85–91, 95, 104–5, 107–8, 111n31; and holes, 132–33
Maudlin, Tim, 14n10, 63–65, 78n3, 136n25, 150–51, 154, 175, 183n16, 185n26, 186n42, 187nn50, 61
McGrath, Matthew, 111n33
Mellor, D. H., 188n65
Mereological Essentialism (ME), 86, 98, 100, 110n12, 118–19, 135n11, 136n32
Mereological Nihilism, 8, 84, 104, 107–8, 108n4
mereology, x, 4, 7–8, 36, 46n5, 83–84, 86, 94–95, 98–99, 110n19, 184n22
Merricks, Trenton, 108n3
metaphysics, embedded conception of, 188n71
monad, 11–12, 15n21, 29–30, 46n6, 83, 113, 138, 163–64. *See also* simple, mereological
Moore, George Edward, 31n9
motion, 2–3, 5–7, 9, 17, 40, 49–54, *53*, 63–71, 76–77, 78n6, 79n23, 80n27, 134, 138, 142, 150–51, 156, 161–62, 184n24, 185n26, 189nn71, 73

Nerlich, Graham, 147, 162
Newton, Isaac, ix, xi, 1–2, 6, 12n1, 17, 46n5, 49, 54, 63–65, 67, 69–71, 77, 79n22, 80nn27, 31, 127, 134n6, 135n8, 138, 144–45, 150, 154, 156–58, 162–64, 169, 176, 183n16, 184n24, 185n34, 186n40, 187n60, 189n73
Ney, Alyssa, 110nn28, 30, 111n31, 182n10, 188n65
Nir, Schlomo, 164
Nolan, Daniel, 13n5, 182n4
nominalism, 185n29
North, Jill, 30n3

Oaklander, L. Nathan, 30n3
objects, ix–xi; composition of and by. *See* composition; four-dimensional, 41, 141–42, 145–47; and holes,

130–31; material or physical, 28, 31n9, 33, 46n3, 85, 91, 96, 108n3, 111n33, 117, 131, 157, 162–66, 173–77, 182n4, 184nn22, 24. *See also* boundaries; materially solid objects; scattered objects

Olson, Eric T., 84–85, 100–103

ontological commitment, principle of, 14n9

ontological economy or simplicity, 5–7, 9, 13n5, 17, 21, 28, 38, 40, 46, 77–78, 102, 114, 130, 132, 134, 136n25, 138, 142, 144, 148–49, 152, 154, 170, 172–77, 179, 181

Pais, Abraham, 184n23

parts, x–xi, 14n16, 182n4. *See also* proper parts; of spacetime, 140, 144; spatiotemporal, 140–42, 146–48, 151, 167, 177–79, 185n28

perdurance, 140–42, 146, 166–67, 179, 183n11, 185n28

persistence. *See* Endurance; perdurance; Spacetime Perdurance (SPER)

Pfeffer, Jeremy I., 164

Poincaré, Henri, 49, 60, 62, 80n31

point: boundary, 115, 135. *See also* boundaries; identity criterion for spatial, 72–73; -sized entity or particle, 10, 184n25. *See also* simple, mereological; of space, xiiin2, 3–4, 13n4, 42, 46, 56–57, 72–73, 162–63, 170–71, 187n60; of spacetime, 47n8, 72–73, 80n34, 138–39, 141–42, 152, 154–55, 163, 165–68, 170–71, 177, 179–80, 182nn3–4, 184n17, 185n32

Pooley, Oliver, 144, 162, 186n42

Presentism, xi, 21, 31n3, 32nn4–8, 171–72

Price, Huw, 30n3

Principle of Sufficient Reason (PSR), 54, 56–58, 78n10, 79n13

proper parts, 7, 9–11; and boundaries. *See* boundaries; and composition. *See* composition; Special Composition

Question (SCQ); definition of concept of, 6–7, 36; and holes, 131–33; and materially solid objects, 10, 90. *See also* parts; and scattered objects, 11, 90–91, 128; and simples, 10–11, 108n3. *See also* simple, mereological; and spatial directional relations, 23–29, 35–37

properties, 4, 10, 30n1, 34, 40, 47n8, 72, 75, 80n35, 151–52

Quine, Willard van Orman, 14n9, 109n6

Rea, Michael, 188n65

Reid, Thomas, 2, 17

relations, 17–18, 92, 101, 184nn24–25; spatial, ix, 3–7, 10–12, 15n21, 17, 20, 33, 38–40, 47n9, 50–63, 65–69, 136n25, 143, 163; spatial directional, ix–x, 7–12, 18–29, *23*, *24*, *27*, 33–39, 42–46, 46n4, 47n8, 49, 53–54, 59–60, 62–63, 69–78, 83, 86, 89, 95, 108, 114, 122–23, *123*, 128, 131–34, 135n19, 137, 142–43, 145–46; spatiotemporal, 47n8, 144, 150, 152, 154, 184n18, 186n38; spatiotemporal directional, 9, 134, 137, 144–49, 151, 154, 167, 175–81, *180*, 185nn28–29; temporal, xi, 21, 30n3, 31n8, 145

Relativity (favored interpretation of (Relativity) or General (GRL) or Special (SRL)), xi, 9, 137–43, 145, 147–51, 154, 156–58, 160–61, 164–67, 169–81, 182n5, 183nn10, 13, 184nn23–24, 185nn26, 34, 186nn42–43, 188nn65, 68, 189n73

Rickles, Dean, 78n9, 188n70

Rosenkrantz, Gary, 15n20

Sauer, Tilman, 187n49

scattered objects, xi, 10–11, 22–25, 46n3, 78, 83, 85, 88–91, 95, 100–104, 106–8, 110nn12, 15, 111n31, 114, 121, 127–30, 132, 134n1, 135n12 sets, 4, 10, 139, 154, 182n4; of spacetime points, 166–67, 177, 182n3; of spatial or spatiotemporal directional relations, 33–39, 43–45, 54, 70, 89, 95, 128–29, 132, 145–48, 154, 177–78

Sider, Theodore, 80n34, 109n6, 111n33, 112n34, 138–39, 182nn4, 10, 188n65

Simons, Peter, 13n5, 132

simple, mereological, 8, 10–12, 19–20, 22–30, 34–35, 37, 46n4, 72, 75, 89–92, 94–95, 107, 108n3, 110n13, 111n33, 138, 168–69. *See also* monad

simplon. *See* simple, mereological

Sklar, Lawrence, 138–39

solid objects. *See* materially solid objects

soul, 12, 15n21, 29–30, 138, 165–66

space: absolute, ix, 1–2, 4–5, 12n1, 50, 60, 63, 65, 139, 162, 176, 189n73. *See also* space, substantivalist; relationalist, ix–xi, 4–12, 17–18, 22, 29, 33, 39–40, 43, 46, 47nn8–9, 49–50, 52–54, 56, 58–63, 67–69, 76–77, 78n1, 80nn31–32, 102, 132–34, 136n25, 137–38, 142–44, 148, 174, 181; substantivalist, ix, 1–7, 9, 13n5, 14n10, 20, 22, 28, 38–46, 46n3, 49–50, 55–57, 59–77, 79n12, 80nn31–32, 91, 96, 101–2, 109n11, 110n15, 114, 116, 130–32, 134, 137–38, 142–44, 148, 151, 163–64

spacetime, x–xi, 9, 78n6; absolute, 139, 184n24, 189n73. *See also* spacetime, substantivalist; bent, curved, expanding, malleable, rippling, undulating, or warped, 156–78, *178*, *180*, 180–81, 183n17, 187n54, 188n70, 189n73. *See also* gravitational waves; and hyperspace, 13n5, 41; manifold, 41, 139, 141–42, 152, 154, 167–70, 178–79, 186n38, 188n63, 189; relationalist, 9, 47n9, 134, 137–38, 142–44, 147–49, 151–55, 161–62, 170–72, 175, 177–81, 183n15, 184n22, 189n71. *See also* Spatiotemporal Relationalism (STR); substantivalist, 47n8, 72–73, 79n12, 80n31, 134, 137–81, 184nn17–24, 188n63, 189n71. *See also* Spatiotemporal Substantivalism (STS)

Spacetime Parity Thesis (STPT), 139–42

Spacetime Perdurance (SPER), 140–42

Spatial Directionalism (SD), 6, 9, 17–29, 37, 39–40, 44–46, 49, 52–54, 56, 58–59, *59*, 62–63, *62*, 69–71, 73–78, 80n31, 86, 95, 102, 108, 109n11, 114, 122, 124–30, 132–34, 138, 145, 151, 181; objection to, 71–77. *See also* relations, spatial directional

spatial directional relations. *See* relations, spatial directional

spatial entities: boundaries of, 113–30; and composition, 83–108, 109n9, 111nn31, 33, 112n34; concept of, 6, 29; holes in, 130–34; kinds of, 9–12, 15n21; and Spatial Directionalism, 17–43; and Spatial Relationalism, 5–7

Spatial Relationalism (SR), ix–xi, 4–8, 14n10; the case against, 49–76; the case for, 39–40, 76–78, 78n1, 132. *See also* space, relationalist

Spatial Substantivalism (SS), x, 1–2, 14n10, 148; the case against, 5, 38–40, 46, 76–78, 143; the case for, 2–6, 40, 49–71. *See also* space, substantivalist

Spatiotemporal Directionalism (STD), 137, 144–48, 154, 177

spatiotemporal entity, 9, 138–42, 144–55, 157, 159–60, 167–68, 173, 181, 184n18, 185n28

Spatiotemporal Relationalism (STR), 144. *See also* spacetime, relationalist

Spatiotemporal Substantivalism (STS), 138, 170, 182n3; assessing the case against, 169–81; the case for, 149–62; reasons to be skeptical about, 162–69. *See also* spacetime, substantivalist

Special Composition Question (SCQ), ix–xi, 8, 11–12, 14n12, 36–37, 77–78, 83–108, 110n20, 111nn31, 33, 112n34, 149. *See also* Conjoining (CON); contact

states of affairs, 10, 18–19, 21, 30n1, 40

Static Spacetime (SST), 139–42, 171, 180, 182n9

string theory, 47n17, 149, 188n63

supersubstantivalism, 182n4

surface. *See* boundaries

Taylor, Edwin F., 145, 156

Temporal Directionalism, 18–22

Temporal Relationalism, xi, 21–22, 30n3, 31n8

Temporal Substantivalism, xi, 20–22, 30n3, 31nn7–8, 138–39, 166, 184n18

Thomasson, Amie L., 108n4

time, xi, 30n3, 31n5, 184n18; A-Theory of, 21, 31nn4, 7; B-Theory of. *See* Eternalism. *See also* Presentism; Temporal Relationalism; Temporal Substantivalism

touching, x–xi, 85, 98–99, 110n15, 111n33

Unger, Peter, 109n4

Uniform Expansion Argument (UE), 60–63, *62*, 134

Uniform Motion Argument (UM), 50–54, *53*, 134

Universalism, 84, 100–101, 103, 109nn5–6

Uzquiano, Gabriel, 111n33

van Cleve, James, 109n6

van Inwagen, Peter, 14n16, 79n13, 84, 91–94, 96–97, 99, 104–5, 107, 108nn2, 3, 109n10, 110nn13, 20–21, 30, 111nn31, 33

Vonnegut, Kurt, 141–42

Wheeler, John Archibald, 145, 156–57

Wilson, Jessica, 186n40, 188n71

Zimmerman, Dean W., 14nn9, 16, 30n3, 183n10

About the Author

H. Scott Hestevold is professor emeritus of philosophy at the University of Alabama. Though he has published essays in moral psychology and in philosophy of religion, his primary research interests have involved problems involving objects, identity, time, and space. One summer, while a graduate student at Brown University, Hestevold taught ethics at both the Tennessee State Prison for Women and the Turney Center for Youthful Offenders. For fourteen years, he taught for the American Academy of Judicial Education, conducting weeklong "Judicial Reasoning" seminars for judges. Few philosophers have taught both ethics to convicts and logic to judges.

www.ingramcontent.com/pod-product-compliance
Lightning Source LLC
Chambersburg PA
CBHW050905300426
44111CB00010B/1382